SAFETY

UNDER CONSTRUCTION

For Construction Workers and Frontline Supervisors

BRION K. HANKS, SAFETY PROFESSIONAL
DAYNA HUBENTHAL, INNOVATION SCIENTIFIC
SCOTT BURR, INNOVATION SCIENTIFIC

Koho Pono, LLC

SAFETY UNDER CONSTRUCTION: For Construction Workers and Frontline Supervisors

Published by Koho Pono, LLC
Clackamas, Oregon USA
http://KohoPono.com

For general information on our other products, please contact our Customer Service Department within the USA visit http://KohoPono.com.

First Print Edition 11september2013

Library of Congress Control Number: 2013916848

ISBN: 978-1-938282-06-5

Manufactured in the United States of America

"If you struggle with taking safety to the next level, then this book is for you. Safety Under Construction goes beyond establishing a solid foundation of where the industry stands, it opens the reader's mind to think differently about safety. I recommend this book to not just leaders but to every person looking to improve safety."

- Charles Martinez, Certified Six Sigma Master Black Belt, Certified Quality Manager

"It is with much appreciation and admiration that I write this... From time to time I was compelled to call and verify your actual thoughts, but with further reading all the answers were disclosed and then some...

In my experience trying to stay alive offshore on oil platforms in Alaska, EVERYONE had significant buy in and we all looked out for each other because the alternative was too huge to ignore...There were many times that our safety briefings, program preparation, and permitting process took as much as 4 times longer than the actual spool replacement or line tie-in. It made no difference what the job was, we trusted each other to fulfill the responsibilities assigned in order to help keep all of us alive. This is not a far fetched exaggeration. It was a way of life for staying alive. I will not sugar coat our results for you. We had accidents over my twenty two years in the heavy equipment construction industry and I saw many things that made me wonder why people would do the things they do. Much of it makes more sense after reading this book. It has been my pleasure to read this life map for survival.

Your book is not really a book but a remarkable journey of discovery and practical application of science for lifelong safety behaviors for any workplace. Pardon my lengthy acknowledgement of your achievements, but I recognize it as a truly heartfelt resolution created by experience and sensibility for real lifesaving methods in heavy construction work and other hazardous work worldwide."

- Bryan T. Mullins, Safety/Compliance Les Schwab Risk Services

"Safety Under Construction is your must-read for this year. I have worked with Innovation Scientific on several projects, and this may be their best work yet. They have delivered a practical approach to make safety the focal point of your business. They have clearly diffused the myth that safety is in the way of profits and performance and show how innovation and safety provide the desired gains.

I recommend this book to leaders and safety professionals not just in construction, but also to anyone that wants to understand the underlying process of how to create and recreate powerful industry scale solutions."

- Dana W. Clarke Sr., Chairman / President / CEO of Applied Innovation Alliance, LLC; Certified TRIZ Specialist and also Certified Organizational Engineer

"I have managed many construction crews and sometimes you face a critical moment where the quality of your field-call affects lives and a successful project outcome. So much is riding on what decision you make in that moment!

Safety Under Construction addresses this important but almost never discussed topic of split-second decision-making. Crew supervisors and managers will appreciate this valuable information.

I've actually heard people in the field say, 'We've been doing it this way for hundreds of years, it's fine' and to the manager in the field, safety can feel like 'Please, just let me get my stuff done.' Safety Under Construction raises the bar to an entirely new level of expectation for worker safety. The whole idea of building safety into process, like quality, is a very important idea in the book. This is a visionary approach to see how safety, profits and process can work together."

- Ophir Orr, Mechanical Engineer, Professional Engineer

"Using systems engineering and innovation principles to improve safety is a new and welcomed approach to safety in the construction industry. I am deeply honored by your invitation to review this monumental work. Safety at work is so critical. My own experience in my first job for Indian Railways left a lasting impression on me. I, as a freshly appointed Assistant Engineer, was at an accident site reviewing the damage to a crossover joint. My deputy was an experienced railway man with over 30 years of service and he was showing me the wear at the joint that had caused the wheel to jump the rail and cause a derailment. In his enthusiasm he used his finger to point at the spot and suddenly the joint snapped to trap his finger and splash blood on our faces. As a 24-year old, I was devastated not only by the accident but the fact that my trusted buddy was subjected to such a cruel event. The prevention principles described in Safety Under Construction, would have spared us that shock, injury and insult."

- Mahantesh Hiremath, Systems Engineer, 2013 ASME Federal Fellow supporting the US House of Representative (Majority) Committee on Science, Space and Technology

"The quotes at the beginning of each chapter are well received. I actually thumbed through the book looking for the quotes to see which ones I liked the most. Also, I like the Safety as a System approach and discussion. The illustrations and the concepts are very thought-provoking."

- Timothy Tangeman, District Safety Manager, Kiewit Infrastructure West Co.

"Rarely does one get to witness the beginning of what could be a historic moment in a field of work. The seeds of concepts sewn in the pages of Safety Under Construction could be the start of significant positive change in the construction safety industry for those who want to achieve great safety and profit performance."

- Harry B. Flotemersch, Certified RedX Master, Structured Innovation Facilitator and Senior Consultant, Shainin, LLC

Dedication

Dayna Hubenthal: To my Dad, Richard D. Hubenthal, a laughing, loving man who died in an accident due to human error. I know first-hand the devastation of a preventable death. And to my second father, David K. Harvell, he taught me the rest of what I needed to know. I love them both deeply.

Scott Burr: To my Uncle Vince - Thank you for the many stories (none of which was ever the same!) about how you lost your finger in an accident. You eventually taught me the value of an honest exchange and that confidence and light-hearted foolishness together can create a serious consequence. To Mom, Dad, Papa, Gragsie, and Auntie Pep - thank you for your love of learning, of solving problems and the belief that tough challenges can be overcome. And finally, to Robert E. Lillo - on the surface you taught me electronics and industrial arts; what you really taught was the value of personal excellence and the importance of detailed preparation and a positive mindset. You have influenced many young minds to be persistent problem-solvers and are an unsung hero. Mr. Lillo, you live in the hearts of those you reached.

Brion K. Hanks: To my parents and brothers and sisters for being there in my life and contributing to the person I have become, thank you one and all. I have intentionally cultivated a passion for safety through learning that it permeates everything we do… a lesson I learned while young.

Industry Leader Acknowledgements

We wish to acknowledge four Kiewit[1] Companies safety professionals who, from a mentoring standpoint, had a very positive and encouraging impact on Brion, one of the authors of this book.

Tim Tangeman is the District Safety Manager for the Northwest Kiewit District and whom Brion worked under as a Project Safety Manager. Through his experience, knowledge, caring power, excellence at engaging craft workers, being strong on accountability and having a great sense of humor, "Tange" led the way for safety improvement year after year and challenged Brion to hone his safety skills constantly.

Brion also worked closely with Troy Marble (District Safety Manager for Kiewit Bridge and Marine). Brion gained deep respect through a close working relationship for Troy and it afforded Brion the opportunity to witness a strong solution-oriented safety leader in action, one who understands that workers are people that must be acknowledged, trained in the key elements of safety and recognized for their efforts.

Rand Magee, Division Safety Manager, walks the talk when it comes to the "Why I Work Safe" programs utilized on Kiewit projects. When Rand visits projects there is a picture of his wife and children affixed to his safety vest. It is a personal reminder for those he comes in contact with that his family is the reason he works safe.

All three men (Tim, Troy and Rand) exemplify safety management motivating in the field. They show workers that management cares and is genuinely committed to safety in the workplace.

And finally, Brion wishes to acknowledge Kiewit District Manager, Shane Petersen, for his strong managerial skills and unwavering efforts to improve safety within

---

[1] Kiewit is a Fortune 500 construction and engineering organization based in Omaha, Nebraska. It is a privately held organization with roots dating back to 1884. It is one of the largest and most highly respected construction contractors in the world and operates through a network of offices in the United States, Canada and abroad. The Kiewit workforce includes 10,300 core staff and 14,400 skilled craft workers. In 2011, Kiewit's revenues exceeded $10 billion - as reported in *Kieways - The magazine of Kiewit Corporation (Quarter 2, 2012 issue).*

AUTHORS NOTE: We cite Kiewit's best practices often throughout the book because they are one of the industry's safety leaders. However, this does not imply a formal endorsement of Kiewit's safety program by the authors and it does not imply a formal endorsement of this book or the authors by Kiewit. This applies to all other companies and people mentioned in this work such as Bechtel, ExxonMobile, etc. unless specifically stated otherwise.

Kiewit. Shane's leadership involves unequaled integrity and commitment that exemplifies "Safety Always".

Their efforts are important in pursuit of achieving the safety perfection we believe is possible, i.e., that all accidents are preventable and that all workers should go home safe each and every day.

Table of Contents

"Don't mistake activity with achievement."

- John Wooden

"With truth, overcome delusion."

- The Dhammapada

Introduction - Why Write About Safety?

We are surprised this issue has gone on for so long.

There is a major problem in the construction industry that has been ignored, hushed-up and pushed to the back burner. And this issue has been putting you at risk and costing your company big bucks as you have all ignored it.

It is the biggest source of confusion on construction projects because, frankly, no one really knows what to do to fix it and you all think you are already doing everything you can do - no more is possible. This book is the beginning of changing how construction safety is viewed especially in relationship to profits, productivity, and schedule.

Big businesses, small businesses, government agencies, and workers are all active in safety today.

In fact, safety is currently the single most acted-upon and yet routinely ignored issue in the construction industry.

We see enormous amounts of activity with relatively minor new achievement. Safety as a discipline needs an upgrade and a reboot industry wide. Misunderstanding, confusion, compromise, and fear are routinely witnessed in the unspoken cold war between production and safety. The result of this delusion (that activity equals achievement) is that people are routinely hurt and killed on the job.

But in the next 10-25 years, you will experience a revolution in how safety and production are addressed on a construction site. It will benefit you if you know about these changes so you can position yourself to keep your job, maybe rise to a higher position, make more money and, all the while, keep you and your co-workers safe.

Technology is one of the drivers of this coming safety change. Many of these technologies are available now, but you have some time before you must jump-in because they have not been fashioned yet to serve your industry. But the changes are coming. Be aware that you must take the lead soon to be well positioned. If you wait too long to see what will happen, it will likely be too late. You have work to accomplish right now!

This book will bring you an immediate competitive advantage by seeing safety and production through new lenses. We will make those lenses clear and magnifying so you know what to do. If you start now, with what you learn in this book and if you make an earnest and effective effort to accomplish an excellent safety record, you will be better positioned to handle the rapid changes that are coming.

These sweeping changes involve advances in knowledge and technology that will make safety more scientific, systematic, and process-oriented. For many companies this means that they will need less people to do the same work. *This means fewer jobs.*

Other more visionary companies, however, can use these changes for unparalleled growth opportunities and many new jobs – growth that has not been seen in decades. You need to prepare yourself for either situation.

Construction sites in the future will be more disciplined, better coordinated and the workers that succeed will be safer, more responsible and technically savvy than workers are expected to be right now. To remain a valuable employee:

- You will need to brush-up some old skills and add some new ones.
- You will need to improve safety for your company.
- You will need to add value to the company's bottom line.
- You will need to be an "obvious fit" into the new technology-based infrastructure.

We are going to help you understand safety in a whole new way by showing how a scientific understanding of safety will help you keep your job, find an easier path to better pay and *keep yourself and other workers safer than ever before.* You can help lead your company through the changes by starting now.

The goal we advocate for safety is 100% zero harm. Oh… and did we mention that this scientific focus on safety will be the central action point for speeding up production and delivering better results on time?

Be skeptical if you wish, but be willing to change your mind. It is our destiny to write this book, save lives and prepare you for the impending changes to come so you remain relevant.

Our Intention

It is our intention to bring an extraordinary new level of mastery to safety in construction.

You will have some time to make up your mind, but if you enjoy your industry, your job, your income, and your delivery of great construction projects, you would be wise to follow our very simple advice. Read this book. Pay careful attention and start growing into safety mastery. These changes will be coming to a construction company near you in the not too distant future. Will you be ready?

Our Qualifications

We each have contributed something very unique to this project, approaching the subject with different roles, experiences, perspectives and knowledge. We have created a breakthrough in understanding safety by seeing the hard realities in construction and by using and reorganizing common and available knowledge in an uncommon way. We also have recognized and integrated the potential role of powerful new technologies to create a new level of safety mastery. The work, however, has only begun.

Brion K. Hanks is a Safety Manager at a major construction company with a passionate dedication to safety. He has over 40 years of experience promoting safety within the construction industry. He is very hands-on and understands the dynamics of construction work and how construction people think. Brion is a world-class safety expert and understands the best practices in his field. He has never stopped searching for better answers to eliminating accidents and has been involved with the implementation of many cutting edge safety practices including the use of a Job Hazard Analysis for all work operations and utilizing exercise-on-the-job to make people safer and less prone to injuries. But Brion's greatest asset is his resolute promotion of safety first and safety always.

Dayna Hubenthal is a Principal at the firm, Innovation Scientific, LLC. She approaches safety from an innovation-based, people-oriented perspective, which is augmented by her research prowess and love of cutting-edge technology. She has a strong understanding of group dynamics, cultural and psychological drivers in groups. Dayna is a professional innovator with a special interest in the intersection of social dynamics, innovation and technology. She is also a premier communicator and has led the communication strategies for this work.

Scott Burr is a Principal at the firm, Innovation Scientific, LLC. He has worked in high-reliability industries and his interest in safety and risk comes from years of hands-on shop floor experience in manufacturing. He is a Six Sigma and Lean Master and also a TRIZ[2] practitioner. He has led several quality initiatives and implementations in various industries. He has solved many complex and entrenched problems including safety issues. Many of these problems were considered "impossible to solve" until he led his teams to solve them. Scott is a Fellow of the American Society of Mechanical Engineers and is dedicated to creating industry scale solutions that provide new opportunities, jobs, and markets.

When Scott and Dayna met Brion, he was already a world-class safety expert. It quickly became apparent that Brion is driven to eliminate the last 5% of 'accidents' that other people within the construction industry find unavoidable. Scott and Dayna formed a collaborative innovation project with Brion as the subject matter expert and focal point of safety passion. Throughout their first conversations, Scott, Brion and Dayna realized a breakthrough was possible as we identified the fundamental industry problems and contradictions. Then Scott and Dayna used inventive tools to lead an analysis of the current industry situation, challenged best practices and then proposed a new and more effective approach to safety in construction. Brion provided an in-depth reality check and push back on ideas that would not fly in construction. This book is one result of that effort and provides a starting point for communicating that paradigm shift to frontline craftspeople, supervisors, and managers.

A Final Thought

In the implementation of safety, character matters. It is up to you to meet the challenge to see safety in new ways and to make your effort sincere. You will need an open mind to get the full benefit of this book. Read this book as if the lives of your friends, family and co-workers depend on it, because they do!

[2] TRIZ is a Russian acronym for Teoriya Resheniya Izobretatelskikh Zadatch or теория решения изобретательских задач, and roughly translates into English as the "Theory of Inventive Problem-Solving".

Bottom line Introduction: ***People continue to get hurt and killed on the job. The new safety goal is 100% zero harm.***

Safety Mantra Introduction: ***Nobody gets hurt near me.***

Take-Away Points: ***Introduction***

- Revolutionary changes are coming to the construction industry.
- Your safety on the job is the single most acted-upon and yet routinely ignored issue in the construction industry.
- There is an unspoken cold war between production and safety, which results in workers being hurt and killed.
- Advances in knowledge and technology will make safety and production more scientific, systematic, and process-oriented.
- Future construction workers need to be more disciplined, safer, and more responsible and technically savvy than current workers.
- Character matters.
- Read this book as if the lives of your family, friends, and co-workers depend on it, because they do!

PART I – CREATING A FOUNDATION

"If you have built castles in the air, your work need not be lost; that is where they should be. Now put the foundations under them."

- Henry David Thoreau, Walden

"I find this a fascinating phenomenon: the ability we have to manipulate ourselves so that the foundation of our beliefs is never shaken."

- Muriel Barbery, The Elegance of the Hedgehog

Because this is an innovation-based book, there are some basic concepts that you need to understand. We introduce these basic concepts in Part I. These ideas will form the foundation of achieving great performance in construction safety.

Also Part I documents the beginning of the paradigm shifts you must make if you want to create an accident-free future. You are here now with your beliefs about safety. When you finish Part I, we envision the start of radical transformation in your understanding about safety and your relationship to it.

To achieve great work you must aim high. You must have a strong belief and must act in accordance with your goal. But even with "castles in the air" the work to "put the foundation under them" is the hard work that makes it real and practical.

High goals, hard work, and achievement offer rewards. So we present a practical look at what is coming in the future of construction so you understand why you must make the effort to think and act differently. You will reap excellent rewards for your effort.

Your journey of understanding safety in a new way begins now.

"It always seems impossible until it's done."

- Nelson Mandela

"There is only one thing that makes a dream impossible to achieve: the fear of failure."

- Paulo Coelho, The Alchemist

Chapter 1 - Safety Defined

It is possible to achieve an accident-free record.

For example, Admiral Rickover and his successors built a fantastic safety record for the United States Naval Reactors program for over 55 years. Because of the core values and safety culture Admiral Rickover initiated, nuclear-powered ships steamed safely more than 128 million miles - equivalent to over 5,000 trips around the Earth - without a reactor accident and with no measurable negative impact on the environment or human health.[3]

"Does that mean no one was ever hurt?" You may ask. "What exactly was their definition of accident-free?" These are good questions and a good place to begin.

The Current Definition of a Construction Accident

For the purpose of this book, we will define a construction accident as "any work-related (external) incident, illness, process, condition or environment that results in medical treatment beyond first aid[4], days away from work, restricted

[3] Statement by Admiral F. L. "Skip" Bowman, U.S. Navy Director, Naval Nuclear Propulsion Program before the House Committee on Science, 29October2003

[4] Medical treatment beyond first aid includes: observation or counseling, diagnostic procedures, including X-ray, blood tests, over-the-counter medicine, Tetanus, cleaning, flushing or soaking wounds, wound coverings, including suture substitutes such as butterfly bandages and Steri-strips, hot/cold treatment, non-rigid support such as ace, non-rigid back belts, etc., temporary immobilization for transport purposes, drilling of nail to relieve sub-ungual hematoma, eye patches, foreign body (FB) removal from eye using only irrigation or swab, simple skin FB removal, finger guards.

work activity, job transfer, loss of consciousness, or death. This includes drug or alcohol abuse and workplace violence."

And we should probably time-bound this statement since the results of accidents sometimes appear later; for example, soft tissue damage or PTSD (Post Traumatic Stress Disorder). So 'accident-free' means none of the above occurs for one year after any incident.

It's Time to Be Realistic

Do you think the claim of zero accidents is a stretch? When we interviewed people for this book, many company owners and safety professionals cautioned us not to claim that we could achieve 100% no-death, no-injury at a productive construction site. Many said the goal was unfeasible and readers would not take us seriously if we claimed this. Therefore, for you skeptics, let's begin the book with this statement, "If you read this book **you will have the tools to help move the number of deaths and injuries to as low a value as is humanly possible.**" Now this is something we can all believe and relate to.

> *"We may not touch a perfectly safe construction site with our hands, but if we choose perfect safety as a goal and we follow that, we will reach our destiny." - Carl Schurz (paraphrased for construction)*

But here is a little warning. We have bigger plans.

In truth, we mean to push our goal much further because we know that much more is possible. By the end of the book, you will be closer to our belief that sometime in the close future a 100% zero accident record is going to be a reality on a productive construction site. And that success will be repeatable. And we hope that many of you will accept the responsibility to take steps to make this happen.

Current State of Accidents in the Construction Industry

In 2012 the 'preliminary' job fatalities statistics are sobering. The U.S. Dept. of Labor and Industries National Census of Fatal Occupational Injuries (people killed on the job) recorded 4,551 fatality accidents in 2009; 4,690 deaths in 2010; and 4,609 people killed in the workplace during 2011[5]. Further, nearly 4 million people suffered an injury in the workplace each year. Think about that: 13,850

[5] Complete results will come out in 2013.

fatality accidents and nearly 12 million people injured during those three years. We categorically state, if one person is injured or dies, that is one too many.

How We Think About "Accidents" Tells Us A Lot

The dictionary definition of an accident is as follows: "An accident is an unfortunate incident that happens *unexpectedly* and *unintentionally*, typically resulting in serious damage or injury." [6]

This definition tells us a lot about the reasons companies still have accidents - it exposes the hidden mindset. In order to help you see this from a more useful vantage point, the authors just need to tweak your point-of-view slightly.

By defining accidents as *unexpected* and *unintentional* incidents a red flag goes up for the professional innovator. The red flag signals us that an unproductive belief system is being used. Accidents happen because there are patterns below the industry's (and worker's and frontline supervisor's) level of awareness:

- If something happens unexpectedly that means we were unable to predict it in advance - that shows a lack of knowledge about the input drivers of safety incidents. To fix this, we need new knowledge.
- If something bad happens unintentionally that means we are performing work (or have tools, processes, and systems) that actually promote harm - there are patterns and priorities out of alignment with our stated goals. This shows we are blind to our triggers and need to understand, expose, and eliminate them.

If something happens unexpectedly or unintentionally then it is not your fault. Right? Well, it depends on how you choose to look at it.

Here is another little tweak to the traditional perspective about accidents. The traditional perspective reduces the responsibility in our minds and hearts. It softens the force of accountability and detracts from uncomfortable facts. This is not a productive way to identify and solve problems.

Great leaders accept responsibility. That's why they are great leaders. Every great leader holds herself/himself deeply accountable. The common person says, "Bad stuff happens occasionally. It's out of my hands. We just have to

---

[6] The word "accident" *Oxford Dictionaries.com* (4september2013) at the following URL http://oxforddictionaries.com/us/definition/american_english/accident?q=accident

accept that this work is dangerous and there is nothing I can do about it. It's always been like this." Great leaders know these rationalizations are flat-out wrong.

By rationalizing thoughts that reduce our sense of responsibility, we allow unhelpful beliefs to gain momentum, which is dangerous. After enough time has elapsed and enough momentum is gained, we don't even question the beliefs anymore; we just think they are 'facts'. The results of acting on these 'facts' over a period of time is that we have created our current situation and will continue to create the same in our future. And that is why you are still producing accidents along with your work.

By now you can see that any sloppiness, confusion or lack of discipline in thinking creates blind spots of understanding. Any false assumptions that we take as fact, leads us off in wrong directions or cuts off future possibilities. This is especially true if all of the people around you believe the same things. If your social and peer groups support your reduced sense of responsibility, sloppy thinking will increase.

Suppose we did discover a more comprehensive theory about accidents. Suppose we could fully predict the "What, When, Where, Why and How" of a potential accident. Could we continue to call them "accidents" and be accurate? Consider this carefully. This question is not a word-game.

New knowledge is rapidly developing in our modern time. Would it be reasonable to predict that in five, ten or fifty years we will know much more about "accidents"? Could we predict that eventually all of the "accidents" which occur today could be understood and prevented in fifty years? If we know that we will understand and prevent accidents tomorrow, then we can at least consider the possibility that we can begin to understand and prevent them today.

In the current situation, we use the word "accident" but our analysis of the situation indicates *there are no accidents.* There is only what we do not know today, the errors we currently make, the consequences of our errors, and what we will know tomorrow. This is a painful reality that if fully felt and grasped can challenge us to become more than we currently are.

Here's a funny thing about expertise. Absolute certainty, decades of experience and deep, unquestioned understanding - all of these things create blind spots. There can be no doubt that these things produce good results, but they can also produce blind spots. It takes energy, insight, thoughtful questioning and the ability to change viewpoints to overcome your own thought momentum and to create a better more accurate world view. And more accurate world views lead to better real-world outcomes.

If your current certainty, experience, understanding and actions still lead to harm on your worksites, it is because you don't understand enough. That's why accidents are *unexpected* and *unintentional.* So one of the best things you can do to stop accidents from happening is to understand more about them.

Throughout our book we will present emerging scientific knowledge, new tools and more productive mindsets so you are better able to achieve your safety goals. Let's call this new point-of-view 'Inventive Safety and Productivity'.

Throughout this book we will remind people that it is important for everyone to understand: safety is not static! In fact, safety needs to be top of mind and fully present. Put another way, safety is always under construction.

Bottom line # 1: ***If we are fully responsible and face our fear and pain, we can realize there are no accidents. There is only what we do not know today, the errors we currently make, the consequences of our errors, and what we will know tomorrow.***

Safety Mantra # 1: ***Safety is always under construction.***

Take-Away Points: ***Chapter 1***

- It is possible to achieve an accident-free record.
- If you are skeptical that an accident free record can be achieved, be ready to change your beliefs, once you see how much can be done to improve safety.
- Great leaders take responsibility and feel deeply accountable to produce positive safety results.
- The word "accident" can mitigate our responsibility and helps us avoid pain and fear.
- If accidents are *unexpected* and *unintentional* then we have a critical mandate and responsibility to improve our knowledge about accidents.
- There are patterns to how accidents occur. If we understand those patterns we can prevent them.

"The purpose is clear. It is safety with solvency. The country is entitled to both."

- Dwight D. Eisenhower

"You have set yourselves a difficult task, but you will succeed if you persevere; and you will find a joy in overcoming obstacles. Remember, no effort that we make to attain something beautiful is ever lost."

- Helen Keller

Chapter 2 - Why You Are Not Safe Yet

So now we know that if you are not completely accident free on your construction sites, it is because you do not yet know enough to reliably produce that goal. The knowledge you currently have is too weak to drive you towards the results you desire. In this chapter we will briefly introduce you to five of the basic concepts that you need to learn. Learning these concepts will help you move out of the hidden mindset and begin to create the results you are aiming for. The five basic concepts are:

- Contradictions
- Countermeasures
- People do weird things
- Poor relationship choices
- Poor commitment/responsibility/resolve

Contradictions

Safety compromises kill and harm people. And yet, compromise seems so practical. It is a treasured tradition in production to make rapid decisions and get the job done. It is also the greatest source of harm and killing on construction sites.

This is the dirty secret and an openly accepted self-deception in construction discussed at the beginning of the book: We tell ourselves that nothing more can be done to improve safety because we can accept a few people dying and maybe more getting hurt but it is unavoidable... right? Face the pain. You know you believe this and you have participated in this grand delusion.

Recognizing it and facing up to this contradiction are the first steps to resolving it.

But let's not go too fast. There is more to this very painful and difficult issue.

So when get-it-done-fast is allowed to prevail over get-it-done-safely, a higher percentage of accidents are likely to occur because safety is compromised with preference to performance. BUT if a safe organization cannot perform to a timeline then that safe company will not be competitive and will not be in business for long. We feel both of these statements are true but in order to make progress, we compromise. We feel we must choose one over the other. Ahh, here we are at the heart of the most critical construction business contradiction. We need to be safe but we need to be competitive, too. And the two are often diametrically opposed. Let's say this even more clearly. Our poor choices and mindsets make us believe that safety is in the way of production.

Unfortunately, many construction leaders believe in this mindset of compromise – quick versus safe – and do not consider the possibility that you can have both quick AND safe.

So here is one of the most important bits of information we will give you in this book. It really is amazing and will help you stand head and shoulders above your peers.

When you observe compromises, excuses, and discussions of trade-offs - especially on very important issues - this is a red flag that must not be ignored. If you look at the problem in the correct way, you can have it all. You will not need to compromise, make excuses, or trade-off one good goal for advancement of another good goal.

It is possible to be safe and finish a project on or ahead of schedule if you learn to have an inventive mindset. Once you learn to think this way, you will be able to help solve seemingly intractable challenges that no one else can even imagine solving.

Identifying and resolving contradictions is an important concept that we are introducing, which will change how you see the world of safety. Remember, this is a brief introduction. We will discuss contradictions in greater detail later in the book.

Contradictions embody a deep understanding and also a deep confusion about the most fundamental challenges in the construction industry. They point to problems and also to opportunities. What are the most fundamental contradictions in construction?

NOTE: *If you would like to make a list, now would be a good time to start it.*

Let's face it, even as we advocate "Safety is first" we also want to have a good wage and work at a successful company. Sometimes it seems that we can only get one thing we want at the expense of the other. Let's look at an example of a contradiction that is very fundamental to construction and begin to create additional clarity of thinking:

1) Construction of hospitals, highways, airports, etc., for jobs and for living, enhances our lives and provides safer conditions for the public-at-large but requires that we face certain inevitable risks to construct them. Our safety is inherently compromised unless we exert additional effort to remain safe during construction.

2) Human life is valuable and families and society depend on everyone's contribution to be happy and whole. Loss of human life is devastating to families, friends, co-workers, peers, companies and society. The trauma affects more people than you could possibly imagine.

Both of the above statements contain high value for humans and thus represent a contradiction in priorities. Which is more important? Is it human life or construction? None of us want to sacrifice our own life for the sake of finishing a construction project. Isn't the instant response that human life is more important? But the outcomes of construction also save lives.

We need the outcomes of construction to make our lives safer but the process of construction is inherently risky. That is a contradiction.

From another viewpoint, what if the risk of losing life appears to be small? Can't I use that tool incorrectly just this once? I'll be extra careful. Can't I ride in that vehicle without a seat belt for only 30 feet? Do I really need a hard hat at the edge of a safety zone?

It is so very easy to slip into compromising. Resolving a contradiction takes more work. You need to be constantly aware of how easy it would be to slip; you must be willing to do the work.

Awareness of risks and then taking a chance on that risk is a *mindset of compromise*, which leads to accidents in line with the statistics of that risk. This is why the last 5% of accidents stubbornly resist elimination. People compromise rather than solve these more difficult problems. This happens more frequently when there is pressure such as falling behind on the project schedule.

One of your most important responsibilities now and in the future is to eliminate the mindset of compromise and develop the skill of identifying and helping to resolve contradictions as a core competency.

By now, you may finally admit to the contradiction between profits and safety. We are bringing it out of the shadows so it can be dealt with for good. At a minimum we want zero recordable injuries and zero lost-time accidents with optimized profits. Please consider that safety gains can come with improved profitability. In fact, in our research we have discovered: once the competition is heated, safety is essential to achieving breakthroughs in profitability.

This is a brief introduction of the concepts of recognizing and dealing with contradictions. Later in the book we will address this subject in great detail. For now, realize that if you are not looking at the world in terms of contradictions, then you will not be able to solve your really tough, entrenched, hidden, or complex problems - and that means you will not be able to achieve a zero harm workplace AND keep your job by helping your company be competitive in the future.

Countermeasures and Solutions

The second important basic concept we will introduce you to is the idea of a countermeasure. Any serious discussion about safety (and why you are not safe yet) should include knowing the difference between countermeasures and solutions.

A *countermeasure* is an implementation of action that holds negative consequences at bay. An example of a countermeasure would be putting up levies around New Orleans to protect the metro area from flooding.

A countermeasure requires constant reinforcement, redundancy, maintenance and vigilance to be effective but never actually fully solves the problem. For example, if the levies are not reinforced, maintained, and watched over vigilantly a big storm could wreak havoc. Sometimes big storms wreak havoc even when the levies are sound, so that is why redundant, alternative countermeasures are in place.

If our countermeasures are good enough, we will never have a problem... right? We'll answer that question with two words: Hurricane Katrina.

The problem with countermeasures is that when dangers are avoided for a while, people change priorities. For example, there were lots of warnings about big storms, but most of the big storms veered off and missed the city. For years, the storms that did hit New Orleans were of insufficient size to overcome the levies and flood the city. Therefore, after a few decades, priorities changed, levy maintenance slipped and vigilance slipped. Even when warnings sounded about Hurricane Katrina many people didn't believe it. These are the reasons why countermeasures are not good enough.

Another problem with countermeasures: if you spend all of your time being vigilant, you will be less productive. Real and lasting safety in construction is difficult to achieve with countermeasures because there are other demands on our attention - notably schedule constraints and making a profit.

Today's culture of safety in the construction industry is based on countermeasures, which requires constant reinforcement, redundancy, maintenance and vigilance. The safest construction companies can achieve a perfect safety record about 95% of the time using countermeasures. If you are achieving 95% harm-free, you are doing a good job on safety... Right? After all, the only time the countermeasures fail is when something *unexpected* or *unintentional* happens.

Other examples of a countermeasure include: [7]

- The back-up electrical system that powers cooling for the nuclear core of the Fukishima Power Plant failed by flooding when an *unexpected* tsunami occurred after the earthquake.
- The inspection of boilers for ongoing service to prevent explosions.

A *solution* is different than a countermeasure in that it solves the original problem very well, once and for all, so that you do not need to deal constantly with it again and again. Here's an example of both a countermeasure and a solution.

EXAMPLE: There was a crew building cables for a jobsite. The brand new manager noticed every worker in the group had small diagonal cuts on their fingers. When he asked them about it, the workers denied there was any problem. Finally, the manager stood out of their view and watched them work. He saw the cable makers spin the wires across their fingers in order to line them up correctly. He also observed them rolling their wrists and massaging their forearms during breaks. They not only had small diagonal cuts on their fingers, they were also experiencing temporary muscle pain that could evolve into what would later be diagnosed as repetitive motion disorder.

Since the workers considered these job hazards to be ordinary parts of their job[8], they were not motivated to change the situation or speak up about the

[7] Some may differ on what constitutes a solution or a countermeasure but we are making this point since we are determined to identify the true risk in safety "solutions". We choose to be very rigorous in our definition.

[8] The cable makers were willing to perform the work without complaint. In their eyes this hardiness distinguished them from "wannabe" cable makers even though the cuts on their fingers hurt.

issues. The manager decided to experience the work himself in order to carefully examine the issues.

He first instigated a countermeasure. He ordered finger cots for the group. The finger cots were thin enough to "feel the wire" which the cable makers required but also supplied enough protection to prevent cuts. This was a very inexpensive fix and he could order them by the boxful so they were always on hand (redundancy and maintenance required). He used the introduction of the finger cots as an opportunity to build trust and help the team see that he was on their side. However, after giving it a fair try, the cable builders did not like the new safety equipment and didn't use the cots.

The manager made repeated visits to their work area (constant reinforcement and vigilance) to assure the fix was used. But they only used it when he was present or reminded them. So, on subsequent visits he continued to address the finger-slice-problem, which continued to build trust. He felt responsible to develop a real solution.

He and another employee eventually designed a rotating tool to align the wires very quickly. The team loved the new tool. Production went way up. Safety and quality was improved. All symptoms of muscle trauma and cut fingers disappeared. Morale rose within the group and the change was readily accepted. These problems never came back. No one ever had to deal with this issue again. The workers, the company, and the customers all won. The implementation of the rotating tool was a *solution* while the finger cots approach was a *countermeasure* because it required constant management vigilance and reinforcement to keep the safety gains in place for the finger cot.[9]

The difference between a solution and a countermeasure is important.

The lines may blur whether an idea is a solution or a countermeasure, but it is the consequences of failure that can help you decide. Here is the major take-away: If you are staking someone's life on a countermeasure, then the vigilance you need in order to keep them safe should be 100% with redundant backup plans.

---

[9] Did you notice the absence of compromise in the solution even though we began with safety first as the objective? If you didn't notice, you may have not yet accepted the possibility of win-win solutions. At the very least, you have not raised your antenna to be on the lookout for such solutions. Keep at it. So much more is possible!

People Do Weird Things

There is another reason perfect safety has not yet been achieved. It's because

people do unsafe things even when they know better. This is the third basic concept we will introduce to you in this chapter.

It is a fact that people don't do the right thing all of the time - even good dependable people. While we can furnish all kinds of statistics regarding this statement, really, it is a gloomy-given which few dispute. To make sense of this truth requires clarity of thinking and a dedication to never mince words on the subject of safety. And we shall not.

Why do people do weird things? We have answers for you. The book is filled with answers. But let's briefly introduce you to something we have written a bit about already. People do weird and unsafe things because the patterns of their thoughts and behaviors can sometimes be very difficult to overcome.

Think about how hard it can be to break a habit, to learn a new language, or to adopt a new viewpoint. The patterns of our thoughts, languages, and viewpoints are entrenched in our brains. Even with Brion's[10] extraordinary communications like, "Do the right thing, even when no one else is looking," he has observed that some people just don't.

Since people don't always do the right thing, we clearly need some help if we are going to improve our safety record.

For example on a construction site, a worker is aware he needs to be safe, but he is also aware that the project is behind schedule. This craft worker needs to climb 6 feet up and yank a board off a wall. It will only take a minute. His boss is waiting with the rest of the crew for him to finish. They want to review their Job Hazard Analysis as a group before they start their next task. He feels pressure to hurry just a bit so he can join the others for their safety planning. Even though he knows better, the craft worker does not take the time to set the ladder solidly on even ground and the angle is not correct. No one is looking. In his world view at this particular moment, he finds it unnecessary to be both safe and fast - the worker chooses 'fast' and compromises 'safe'. Why does he do that when he knows better?

---

[10] Brion K. Hanks, one of the authors of this book.

Fear and Pain

Two major factors drive unsafe behavior on the job. They are both forms of hidden compromise. To make progress, we need to bring them into the light and address them directly. The two psychological drivers are: (1) Fear and (2) Pain. [11]

Let's start with fear. For the purpose of safety, there is productive fear and unproductive fear. A productive fear indicates real personal limits, what may or may not work, or what may put someone in harm's way. Honoring a productive fear always enhances safety.

Unproductive fears introduce risk. Fear of losing one's job, focusing on one aspect of the job to get ahead, decision-making under stress, or self-esteem fears based on performance or lack thereof are all examples of unproductive fears.

Fear is static that prevents me from hearing myself. - Samuel Butler

Consider unhealthy psychological forms of fear and behavior such as paranoia, sociopathic, or severe anti-social behavior. Edward Deming, the great Quality Guru, emphasized the need for driving fear and negative behaviors out of an organization in order to achieve new levels of performance. This is especially true when it comes to achieving new levels of safety performance.

You can do much about eliminating unproductive fear and thus the inherent risk it brings to safety. The contradictions related to fear may be difficult to express or to conquer because most people find it difficult to face fear and may not admit they have a perceivable weakness. Fears are very personal; expressing them may hurt your reputation - also a contradiction. Let's realize in the current situation that fear generates emotionally charged safety problems that are contradictory.

Now let's briefly look at pain. Physical pain can be a safety risk because pain steals attention. Physical comfort and sustained attention are both very

---

[11] Because we are innovators, we viewed hidden behavioral drivers from different points of view. Fear and Pain remained relevant no matter what view point we took. On the subject of safety, we looked through the lens of W. Edwards Deming's ideas on Quality, took a Pavlovian view of stimulus/response conditioning, the Instinct Theory of Motivation by Williams James, Aristotle's ideas about all living organism's desire to pursue or avoid as basic motivation. We also considered the work of Sigmund Freud, NLP and NAC, marketing theories and other points of view. No matter what vantage point we took, when you drill down into safety to discover the hidden motivations, the most relevant are pain and fear.

important to safety. Yet some jobs require intense exertion that may lead to pain.

Pain can also be very subtle as a cause of safety risk. For some it is "painful" to walk just an extra few feet to put on safety glasses or secure a hard hat. Change may be painful to some, making it difficult to adopt new safety measures. Some workers associate "waiting" with pain and they have poor patience. There are personality issues and psychological profiles that find it painful to relate to others except in their own specific modality. This may cause resistance to training and new measures or new people. Although this type of pain is really more psychological, it has a very real affect in the workplace as risk.

Fear and pain drive decision-making, putting a person in contradiction with competing priorities. In the current situation, we can clearly see that physical and psychological fear and pain must be addressed if we are to create a safer work environment.

Ignorance and Assumptions

Everyone intends to go home safe after a good day's work. But the way people run their brains will often undermine this good intention. Ignorance and assumptions kill people. Here are a few basic patterns you may have seen:

- Excuses
- Russian roulette
- Foolish and negative leadership
- Not thinking

Excuses - Excuses are a pattern of ignorance and a desire to avoid responsibility (based on fear). We all hear excuses on every project.

"Nobody told me…" is an excuse in which many people indulge. It is a great way to deflect responsibility - and is totally unacceptable. We often use excuses for not being observant, for not thinking clearly, for avoiding responsibility or for not knowing. Excuses can be bold (such as "Nobody told me") or excuses can be subtle (such as believing *unexpected* and *unintentional* safety lapses are not our fault). Excuses need to be eliminated completely from the construction site.

A woodworker used her own extension cord because it was handy rather than going to the equipment shed and getting the proper cord with a grounding prong. When the worker was "caught" the worker first made the excuse, "It was handy so I used it." The safety manager told her that was against the rules so her next excuse was a deflection technique, "Where does it state that rule?" She was shown the rule so her final excuse was, "Well, no one told me."

Excuses are unproductive. Excuses lead to accidents. Most good safety programs attempt to eliminate all excuses.

Russian roulette - Russian roulette is a gambling game where the players put a bullet in a gun, spin the cylinder, place the nozzle to their temple and pull the trigger. If it goes off, you die. If it does not go off, you win that round. Gambling your safety to gain a few minutes of delivery time is a seductive form of "playing the odds" and is a form of ignorance.

- It is only 20 feet. I don't need my hardhat. I will be back in a few seconds.
- My safety glasses keep fogging. I can take them off for just this operation.
- It is only 7 feet up. I don't need to tie off. It will be just a second.

Statistical risks may happen 1 in 100 times, so some people can go years without ever having an accident, being sloppy and breaking rules. Years of playing Russian roulette with safety can create false confidence. The worst part is that your attitude can spread to others and thus YOU can be causing someone else to reap the dangers you disregard and spread.

Negative Leadership - We know this type of worker and sometimes we believe in them. They either have charisma and communicate well or they can be a bully. They are often skilled at their job. They have a rationale for why breaking the rules is okay. Their disregard is a congruent and mighty belief and it draws people to them. A little core group of people can develop with a hostile disregard for safety and disrespect for the intelligence of those outside their group.

Despite their talents, in the end, these negative leaders do not improve productively within their company because part of their unconscious goal is to be great by my making others look stupid. Rather than delivering a solution that works without compromise, these workers are interested in positional power and maintaining the status quo. This is a negative leader.

Negative leaders encourage complicit behavior. They mock management and other workers and crews. They encourage conflict. They could be a great company asset if only they could be trained to be a productive leader.

Not Thinking - Sometimes we need to do something quickly and we move to make that happen. In a rush and with a narrow focus, safety can be forgotten.

For example, a manager drove onto the site in his truck and parked it behind a large evacuator close to the supervisor he wanted to speak to. This was not in the designated safety area, but he was a high-level manager and he only visited the site once in a while, and he was in a hurry, and he only needed one bit of information which wouldn't take long to gather. He did not take the time to ask, "What are the site safety protocols?" Instead he parked his vehicle, exited it (with no PPE) and walked over to the field supervisor he needed to speak with.

In just a few seconds, the evacuator backed up and his vehicle was crushed. Thank goodness he wasn't still in his vehicle!

All versions of ignorance and assumptions are costly. To develop solutions for these major safety issues that exist today, we need to deal with people's dangerous thinking, habits, behaviors, and beliefs. We discuss this in detail later in the book.

Poor Relationship Choices

Safety requires cooperation, mutual respect, process understanding and problem-solving skills.

You are not yet safe because many relationships in construction are based on opposition and not upon mutual respect and cooperation. This is the fourth basic concept we'll talk about in this chapter. Many people in construction dismiss this subject. They do not realize how pervasive relationship choices are intertwined with the unsafe hidden mindset. So, set any instinctive skepticism aside and consider this with a fresh viewpoint.

Briefly and silently consider your personal relationship to OSHA, safety standards, and the diverse range of safety relationships in construction. If you are thinking...

- OSHA is costing our company,
- Safety inspectors are idiots,
- I better shape up when the safety manager comes around,

... then you can improve your company's bottom line and safety performance by building better relationships and choosing a solution-oriented attitude.

Obstructive conflicts hurt safety performance and profits. Mutual respect and cooperation improves safety performance and profits. Industry leaders such as ExxonMobile Development list "a partnership approach to safety" as one of their top safety management principles. Everyone wins with this approach: the contractor, the company, and the worker. The basics of this approach are cooperation, teamwork, alignment, sharing, flexibility, and mutual respect.[12]

This is a great and underutilized opportunity that permeates construction. This is not as obvious an answer as it seems as you will see later in the book.

Poor Commitment / Responsibility / Resolve

Great leaders (and great innovators) have a number of behaviors, characteristics, and mindsets in common. Although we discuss this in great detail later, there are a few ideas that are so basic they must be introduced early in the book.

If you are going to achieve safety perfection AND simultaneously improve your company's profits, then you will have to be realistic. You have to squarely look the current situation in the face. You cannot avoid ugly hidden beliefs or shrink from your fears and pain.

You also cannot bow to the pressure of the rest of the industry (and your company and your peers) who claim that what you are attempting to do is impossible. You must face the truth that the current situation is dismal and there is no known solution for what you are trying to do. There is no path into the future you want; you will need to blaze your own trail. That takes courage and commitment.

Also, you must demonstrate a passionate focus, dedication, and absolute certainty that the goal (zero harm) is achievable and must be achieved even if you do not yet know exactly how you are going to achieve it.

You must take responsibility for the goal.

And finally, you must persist until you prevail.

Bottom line # 2: You are not yet safe on a construction project, because your safety knowledge is insufficient to produce that goal reliably.

---

[12] Jack Toellner, P.E., CSP, MPH; Senior Technical Advisor at Safety ExxonMobil Development Company in a speech to the American Society of Safety Engineers on the topic of Safety in the Petroleum Industry on 23september2010.

Safety Mantra #2: ***Risk comes from not knowing what you are doing.***

Take-Away Points: ***Chapter 2***

- The big five issues that prevent you from being safe and contribute to accidents:
    1. Unsolved safety contradictions
    2. Believing countermeasures and compromises are solutions
    3. The weird things people do
    4. Aggression in relationships
    5. Ineffective commitment and responsibility for safety
- The big contradiction in construction is get-it-done-quick versus get-it-done-safe.
- Compromise is a major root cause of accidents, fatalities and safety program failure.
- Two psychological drivers of unsafe behavior on the job are both forms of hidden compromise: they are (1) Fear and (2) Pain.
- Ignorance and assumptions cause harm and kill; some indicators are excuses, playing Russian roulette with safety, foolish and negative leadership, and unthinking actions.
- Safety contradictions point to productivity problems. These contradictions have always contributed to accidents.
- If you are staking someone's life on a countermeasure, then the vigilance you need in order to keep them safe should be 100% with redundant backup plans.
- Relationships based on cooperation, mutual respect, process understanding and problem-solving skills deliver positive safety results.
- You must demonstrate a passionate focus, dedication, and absolute certainty that zero harm is achievable and must be achieved even if you do not yet know exactly how you are going to achieve it.

"Is the system going to flatten you out and deny you your humanity, or are you going to be able to make use of the system to the attainment of human purposes?"

- Joseph Campbell

"Nothing in life is to be feared, it is only to be understood. Now is the time to understand more, so that we may fear less."

- Marie Curie

Chapter 3 – What's Coming and How to Prepare Yourself

How is the construction industry going to change in the next decade? What can you do to assure you still have a place in that new future? These questions are probably not yet on your mind. In fact, big change probably does not even appear on the horizon. But big change is coming and some of the changes will put many construction workers out of a job.

All throughout this book we write that you need to add value in this soon to be changing world of construction; but until you understand why you need to make these changes, you may be reluctant to act. So, this chapter introduces the basics of what is changing and why.

The authors' ideal goal for business is that every project and every department would use inventive problem-solving, statistical methods, process focus and science as the path to *"Safety and Quality built-in, with cost and performance as prime considerations and meaningful work for all."*

Some companies today may already use these methods during the construction design stage in varying degrees but not as a comprehensive strategy to eliminate safety issues. Here is a little peek into the future through five lenses.

- Technology
- Safety as a System
- Safety as a Science (Quality Manufacturing)
- Standards and Certifications
- Attitudes

Technology

Technology will be everywhere. Technology will be smarter. You will need to interact with technology. There are several high-leverage technologies that have solid potential to make the construction site very safe and more productive at the same time.

Low cost semi-conductor chips, sensors, cameras and microprocessors will make it easy to have smart technology all throughout the job site to support safety. Onsite equipment, Personal Protective Equipment (PPE), your smart phone, and wearable technologies all have the potential to make you safer and more informed on the job.

Construction jobs are complex and dynamic and there are plans to create technologies that can monitor and support a safer and more productive site.

Data and Metrics - There is going to be a lot more data collected and analyzed to improve safety and production. Streams of data will be easy to collect using the inexpensive sensors and other devices. There are specific key factors that contribute to safety success. You may be surprised but these same factors will also support improved production. Much of the data will be collected, analyzed and acted-upon in real time. That will be a part of your job. In the future, almost everything you do will be monitored in some way and data will be collected and used.

Simulation Models, Risk Analysis, and Other Decision-Making Tools - The Workplace Hazard Analysis (dictated by OSHA[13]) will be supported by computer simulations. When conditions change like weather or site layout, simulations will help you understand your new risks and hazards in real time. Also, there will be much better thinking strategies used on a construction site to analyze risk. New tools will predict failures before they happen. In the future, you will be expected to interact with technology to avoid injuries to yourself and others.

Lean Six Sigma - The basic premises of Lean Six Sigma are to eliminate waste, reduced process variation and to standardize processes. If you help do this, you will be safer and more productive. Earning a Lean Six Sigma certification will help your future career.[14]

---

[13] OSHA regulation: 1910.132(d)(2) - https://www.osha.gov/pls/oshaweb/owadisp.show_document?p_table=standards&p_id=9777 (accessed 2aug2013) and eTools S&H Management Systems webpage.

[14] To learn from the best, here are the top 5 in our opinion: ASQ a professional quality org. (http://asq.org/sixsigma/); SME a manufacturing professional org. (http://www.sme.org/Search.aspx?searchtype=sme&searchtext=Lean%20Six%20Sigma&searchfor=all&orderby=id&orderdirection=ascending); Toyota Productions System training (find a location near you); Juran Institute (http://www.juran.com/), and BMG (http://www.bmgi.com/training/certification-information).

Safety as a System

In the next decade, safety will be systematically coordinated. There will be a lot of technology 'talking' to each other and keeping track of you and equipment and tools and materials.

Sensors, chips, microprocessors, cell phones and wearable technology will monitor and support safety. But there will also be a system that takes all that data and manages it automatically. The system will be very smart, fast, and pervasive.

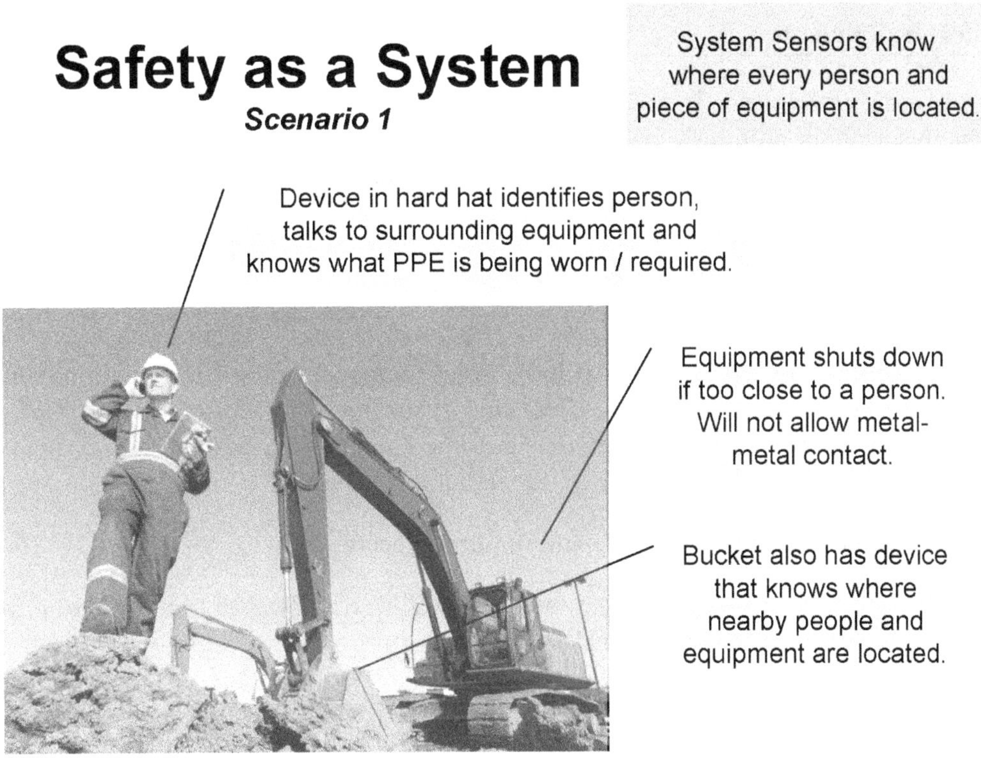

Software Could Track the Movement of People and Large Equipment In Real Time

The concept behind safety as a system is to coordinate and leverage all of the data to enable better preventative and real-time decisions.

- Imagine being able to stop a truck automatically that is backing up and does not see that someone is behind it.
- Imagine having a visual of the entire job site informing you of "hot-zones" where risk is the highest.

- Imagine that you are reminded to take a real-time training using your cell phone just before you operate a piece of equipment that was recently modified.

Some important benefits of safety as a system include:

- Better site safety, crew safety, and personal safety in real time;
- Better on-time performance of your job;
- Better safety outcomes and improved productivity and higher quality.

Safety as a Science

Manufacturing has a lot in common with construction. Manufacturing used to have lots of accidents and early manufacturing companies felt that doing business was inherently risky to employees. People died; it was part of the risks of working.

To a large extent, the authors believe the statement above is the current mindset in construction. Craft workers think, "You will just have to deal with that if you want a job." In fact, some feel it is a distinguishing aspect of their work.

At the turn of the last century (1900), even the most responsible manufacturing companies optimized (traded-off) safety against productivity. *[Sound familiar?]* Yet, we also know that today manufacturing has become a much safer place. Many companies now enjoy a zero accident workplace. How did they do it?

They discovered a simple, valuable and important secret:

A focus on process improves quality, productivity and safety. When the process is understood scientifically then quality, productivity and safety soar.

But it is true that construction is not as standardized, as controlled, or as well understood as manufacturing and, thus, is less able to make the claim that this strategy will work.

But then again whose fault is that?

Just like manufacturing, once construction processes are better understood then there will be a whole new level of performance in safety, productivity and quality. But here is advice for you workers and leaders who are not convinced:

It is not necessary to change. Survival is not mandatory. - W. Edwards Deming

All three authors have witnessed first-hand the sweeping technological and social changes that came to the manufacturing industry. The move from a crafts model of production to a process model changed everything. It is now the construction industry's turn but with a twist.

One of the authors personally ushered in the changes in manufacturing at critical competitive junctures. At that time it was popular for management and the production force to think that zero defect quality was a joke – a big joke. With scrap rates as high as 30%, many did not take automation or the quality revolution seriously since they had a large market share. And there was plenty of misconceptions and confusion about quality during early implementations. But ultimately, the scientific focus on quality and the need for acceptance of responsibility protected the wise. The foolish misinterpreted the change, pointed fingers, lost market share, bled jobs and lost relevance. History now records that many leading companies were destroyed by competitors that delivered better quality.

Among the winners, some modern companies have operated without product defects for several years running. In the Six Sigma tradition of quality, some manufacturer's processes are so unlikely to create a defect that you are more likely to win a $1,000,000 jackpot in the lottery than for them to produce a defect.[15]

Think of accidents as defects. Hmm...?

Our Intention is Safety Mastery

It is our intention to bring an extraordinary new level of mastery to safety in construction just as was accomplished in the manufacturing industry for quality, but with less confusion, fear, and pain - and perhaps with more jobs and not less.

Standards and Certifications

The marketplace is global. So an interesting way to view the next decade is through the lens of standards.

If big sweeping changes are coming and process is increasingly important then jobs will likely be reduced just as it occurred in manufacturing. This is the classic employer perspective. So how do you create a strong competitive advantage for yourself on the job and a strong argument for increased pay while improving safety for yourself and others? The answer is use standards and certifications - use the standards and certified knowledge that are growing rapidly or are becoming global.

---

[15] The odds of winning a $1,000,000 jackpot are 1 in 5,153,633 (per the Powerball Website on 05august2013). This compares to a 6.57 sigma level process using the 1.5 sigma shift suggested by Motorola. Many processes in manufacturing are now this reliable.

Consider the following standards and how you can leverage them to improve safety and be more valuable to your employer while protecting your job:

- OSHA: Be an expert in applicable OSHA standards.
- LEEDS: Become certified in a leading sustainability standard.
- Systems Engineering: Join a systems engineering organization and learn systems thinking. INCOSE is a leading industry group for systems engineering.
- Project Management: Become certified to manage safety initiative projects.
- Unions: Receive tool and safety training from the unions; ask for certifications.
- Lean Six Sigma: Learn to be a Lean Six Sigma professional so you can improve processes, trainings, and safety outcomes.

These are just a few samples. It is your responsibility to find how you will fit into the future of construction and gain an advantage by learning new skills to keep people safe on the job. Defend your job by utilizing standards and certifications.

Attitudes

You will need a new attitude towards your work in the coming decade with regard to safety. The belief that safety interferes with productivity and profit is actually absurd.

Today it is normal and ordinary to think that nothing more can be done to improve safety. This is an outdated and closed-minded thinking process – so let it go. The foundation of this error is rooted in a long tradition in construction that focuses on production first. Later in the book you will discover how this is an attitude of compromise and we will give you tools and tell you how to reduce and eliminate compromises in your safety performance.

While we will discuss many other important attitudes here we will identify four:

- You must have a mindset of "no compromise" with regard to safety.
- You must accept responsibility to improve safety.
- You must be more process-oriented.
- You must think before you act.

Bottom line # 3: Processes and projects will have safety and quality built-in, with cost and performance as prime considerations and meaningful work for all.

Safety Mantra #3: ***Safety solutions must be win-win.***

Take-Away Points: ***Chapter 3***

- "Safety" will be more scientific, systematic and technology-rich.
- Interacting with technologies will make you safer and more informed on the job.
- Learning new skills and standards will keep you relevant.
- A process focus improves your understanding of safety, productivity and quality.
- You will need to have an attitude of "no safety compromises".
- You must accept responsibility to improve safety.
- You must become process-oriented.
- You must take time to think before you act.

PART II – LESSONS AND BEST PRACTICES

"There are no limits. There are only plateaus, and you must not stay there, you must go beyond them."

– Bruce Lee

"The future is already here – it's just not evenly distributed."

- William Gibson, The Economist, December 4, 2003

There is much to learn from our past mistakes and also from observation of the best safety performances in the construction industry. But learning occurs in the context of your mindset and level of awareness. So any lessons we take from the past or from the best must be framed from a point of view. Also if we hold ourselves to a higher standard than those that came before us we will see new value in old lessons and best practices.

What was considered impractical a decade ago, may be possible today if we identify the opportunities.

With the accelerating pace of new knowledge creation and with science evolving so rapidly, we can legitimately say that what we think is impossible today may already be developed and waiting for us to apply to safety in construction.

In our analysis of the lessons from the past and the best practices in construction safety, we see new opportunity for improving safety with proven methods and technologies that the industry has not applied, but has been successfully used in similar industries.

If the future is already here and the knowledge exists to dramatically improve safety, but you are not on the "distribution list", then read Part II and the rest of this book carefully. View this content with the mindset that safety perfection is possible, that solutions do exist, and that learning deeper lessons requires a higher standard of expectation for safety. We will reveal some very practical information that will help you improve your safety practices.

By exploiting the right opportunities, safety perfection is not only possible in the future, it may perhaps be attainable now.

"If you do not know how to ask the right question, you discover nothing."

- Dr. W. Edwards Deming

"We are products of our past, but we don't have to be prisoners of it."

- Rick Warren, The Purpose Driven Life

Chapter 4 – Analysis of Safety Flaws and Failures

Over the years, improvements in regulations, technology, and understanding have had a strong positive impact on safety in the workplace. This has resulted in better conditions for those who perform construction work. We predict that developments not yet discovered will continue to create new ways to monitor and improve safety in the years ahead.

But some things do not change. The following are examples of safety sins, flaws and failures over the past hundred plus years. The picture we paint is not pretty but it makes our point about the on-going and persistent problem of fatality accidents in all workplaces.

EXAMPLE ONE: Major Mine Disasters in the United States, during the early years of the 20th Century[16]

The year was 1907. An explosion at the Monongah Coal Mine in West Virginia took the lives of at least three hundred and sixty two people. Many were young boys. A grave digger states that he believes the real death toll was near six hundred and twenty people. Both numbers today would be shocking and there would be demonstrable public outrage. Regardless, two hundred and fifty widows were made and one thousand children lost fathers that fateful day – a day that remains the worst single mine disaster in the United States.

Just two years later, in 1909 at the Cherry Mine in Cherry Hill, Illinois, nearly five hundred men and boys, and three dozen mules, were working in the mine. Because of an electrical outage that started earlier in the week, the miners were using kerosene lanterns and torches during their work shifts. A coal-car was carrying hay for the

[16] Mine disaster information was presented at the United States Department of Labor Mine Safety and Health Administration website - http://www.msha.gov/DISASTER/MONONGAH/MONON1.asp. We accessed it on 12mar2013.

mules and brushed by a lantern catching fire. It spread to timber supports for the shafts... When it was over, two hundred and fifty nine people had lost their lives as a result of the fire in that mine according to the Mine Safety and Health Administration (MSHA).

While the above two notable mine disasters stand out, between 1907 and 1909 there were actually forty nine mine disasters that took a total of 1,773 lives. And this was the human toll just in the mining industry.

In the day, "crude" is an understatement when it came to mine safety. Imagine that using canaries or "waltzing" mice as gas monitors was the way to tell whether or not there was methane gas or other flammable gases in the mines. Oxygen deficiency was another hazard. If a canary remained alive or the waltzing mouse kept "waltzing" then gas levels were low enough that miners continued to work.

It wouldn't be until the 1970s that more effective gas monitors were developed and put into use with the result being a dramatic improvement over the crude methods of detection for dangerous gases.

NOTE: *This one improvement represents a deeper understanding of the business of safety in mines. But as an abstracted inventive idea, any scientific understanding of safety will improve safety results. We will keep this in mind in our analysis of the history of safety and how it may influence the future of safety.*

What also plagued coal miners is slow death by way of "black lung", which has shortened the lives of hundreds of thousands of workers, and continues to affect the health of miners adversely to this day.

In those early times of mining there was no real training or understanding of the most serious dangers to workers. It is not hard to argue that the industry was too slow to bring about better safety methods.

Typically, fingers were pointed at management for the failures due in large part because of focus on production and profits over miner safety - probably because they really did believe nothing more could be done. This shows the traditional hidden mindset.

In 1891, the U.S. Congress passed the first federal statute to govern mine safety. Its main accomplishment was to establish minimum ventilation requirements in underground coal mines and also to prohibit employing children less than 12 years of age. Imagine children working in such conditions.

Other laws came along that contributed to improving mine safety but it took the U.S. Congress passing the Mine Safety and Health Administration (MSHA) Act in 1977 to begin to produce dramatic results regarding mine safety. Mining fatalities dropped sharply following the implementation of the Mine Act from 272 in 1977 to 34 in 2009. And yet, fatality accidents continue to happen in this industry.

The success of this law reminds us about a rule of thumb in Structured Innovation, which states that 80% of the problems anyone faces have been solved in another industry or field of study. We are also making a personal note to use this principle to understand safety improvement as we progress through this analysis. Consider making notes about what problems others have solved and see if they apply to your situation.

Many of the current best practices come from a morphing of OSHA, MSHA, and industry best practices.

Because of advances in regulations, research, technology, and preventative programs, the number of mine-related deaths has dramatically declined. In examining the history of mines, why did it take laws to get companies, managers and individuals to begin to do the right thing? How is this similar to your project management priorities and business models? When thinking about young miners going down into a mine for the first time, can you imagine what they felt when they first breathed the diminished air quality? Why are these questions avoided even by workers today? What priorities must be set by management, labor, and government to ensure constant vigilance and increasing safety?

Despite advances, the death rate in the mining industry is too high even today.

- 2010: Twenty-nine workers were killed in an explosion at the Upper Big Branch Mine of the Massey Energy Company in Montcoal, West Virginia.
- 2007: Six miners and three rescuers were killed in collapses at the Crandall Canyon mine in Huntington, Utah, owned by a subsidiary of Genwal and Murray Energy Corporation.
- 2006: Twelve people killed in a methane explosion at the Sago Mine in Tallmansville, West Virginia owned by International Mines Corporation.
- 2006: Five people killed in an explosion at the Darby Mine No. 1 in Holmes Mill, Kentucky; owned by Kentucky Darby, LLC.
- 2001: Thirteen people killed in explosions at a Jim Walter Resources Inc. mine in Tuscaloosa County, Brookwood, Alabama.

EXAMPLE TWO: Hoover Dam Construction between 1931 and 1935

The building of Hoover Dam was one of the greatest construction feats of its day. It was built for irrigation water, hydroelectric power, and jobs. It is a National Historic Landmark and according to the American Society of Civil Engineers, Hoover Dam is one of seven Modern Civil Engineering Wonders in America.

Because of it's history and the scenic beauty of the surrounding canyons, the dam itself, and the ability to drive across the top of the dam makes this a popular tourist attraction. Many, many years ago a Hanks relative[17] said, "Always praise the bridge that carries you safely over" because many sacrificed their lives for this convenience. Hoover Dam is such a bridge that should be praised by all travelers crossing over, and here's why.

In December 1922, Reclamation employees Harold Connelly and J.G. Tierney fell into the river and drowned while looking for a location to build the dam. Tierney's death is the first death associated with the building of Hoover Dam. In an awful irony, exactly 13 years later, Mr. Tierney's son, Patrick W. Tierney fell from the intake tower. He was the last worker killed during the construction of the dam - a heavy price for one family to pay. [18]

Between those two deaths about 110 deaths are associated with the construction of the dam. Does that number seem high for four years of construction? Yes it does. What is not as well known, but should be, is the number of deaths not included in the official fatalities statistics that were deaths recorded as a result of "pneumonia". That's right - pneumonia.

Workers on the dam alleged that there was a cover-up regarding deaths due to carbon monoxide poisoning. If it was called carbon monoxide poisoning, the company had to pay compensation claims. If it was called pneumonia, however, the company did not have to pay compensation claims. In total, the number of "pneumonia" deaths was 42 and the number of recorded carbon monoxide poisoning deaths was zero. Most of the workers for the dam construction lived in Boulder City and it is interesting to note that during the construction of the dam there were no deaths of non-workers in the city due to pneumonia. This is powerful data that raises suspicion. Observers have noted it is highly likely there was suppression of the truth.[19]

In examining the history of "accidents" during the construction of the Hoover Dam project do we see a familiar pattern in human behavior? If we are to frame this in the words of a contradiction, it appears that those who have responsibility in the face of an accident may not actually accept responsibility, but try to frame the story. Cover-ups are human behaviors that belong to an unsafe

---

[17] Brion K. Hanks is one of the authors of this book.

[18] *Fatalities at Hoover Dam.* Essay. U.S. Department of the Interior, Bureau of Reclamation, Lower Colorado Region website http://www.usbr.gov/lc/hooverdam/History/essays/fatal.html (Retrieved 31august2013)

[19] Hiltzik, Michael A. (2010). *Colossus: Hoover Dam and the Making of the American Century*. New York: Free Press. ISBN 978-1-4165-3216-3 (pages 284-286).

mindset. In this example and many others, a "get it done at any cost" mindset combined with weak safety habits created an unsafe jobsite. Add to that a tendency to ignore the reality of the outcomes and we see a deadly mix where performance/production was likely the default priority over safety.

In retrospect, we can see that this is neither preventative nor inventive thinking. In making your notes, consider how you would respond if you knew in advance that 152 people would die on your project. Consider also that your country is watching your performance. What measures would you take to prevent deaths and move forward? What types of solutions might you employ to create better safety in the face of unknown risks?

Billboards located around Hoover Dam reminded workers that "Death is so permanent." We agree. No one can argue that far too many workers died during the construction of Hoover Dam.

EXAMPLE THREE: Canary Islands – Two 747 Planes Collide on a Runway[20]

The year was 1977 when several circumstances put two Boeing 747 planes at the wrong place at the same time. The deadly result was that five hundred and eighty three people tragically lost their lives. Neither plane had been scheduled to be at the Tenerife's Los Rodeos Airport in Spain's Canary Islands.

What precipitated the worst aviation disaster in history? Many small factors contributed and if any of the following events had not occurred this would not be a topic of discussion: A bombing one hundred miles away at another airport; it was Sunday and only 2 air traffic controllers were on duty and they were not accustomed to handling that much traffic; there were not enough fuel truck people, places to park the planes, or facilities for all of incoming people. Also, the Pan Am plane was late departing from Los Angeles; the KLM had taken time to refuel which allowed fog to deepen. KLM topped off the fuel, which meant it must go faster on take-off. In the meantime, the Pam Am was unable to get around the KLM when it was refueling otherwise it would not have been there when the KLM was taking off. The Pan Am missed the taxiway it was instructed to take; the Pan Am and the control tower were speaking at the same time and therefore the KLM did not hear that it was NOT cleared for takeoff. If the visibility had only been a little bit better the planes would have seen each other sooner.

And perhaps the most important part of these events culminating with the two 747s on the same runway at the same time was the pilot of the KLM began takeoff and his

---

20 NOVA. "*The Deadliest Plane Crash*", Aired October 17, 2006 on PBS. Quotes were on the NOVA website, http://www.pbs.org/wgbh/nova/space/deadliest-plane-crash.html (accessed 31august2013).

co-pilot tells him that they have not been cleared for takeoff. Now the pilot is irritated, probably embarrassed and thus emotion comes into play. The pilot then begins takeoff again and because no one wants to embarrass him again, no one stops him. They die because the cockpit crew silently watched him start the takeoff without clearance during reduced visibility conditions... and the rest is tragic history. The unimaginable happened and we are left to figure out why.

Are there lessons to learn from this disaster that can inform us today? From the viewpoint of psychological inertia: reputation, emotion and denial can play an enormous role in the safety or risk that we face.

This pattern of denial and emotion was also present in the Titanic disaster and more recently in the grounding and crash of the Italian cruise liner Costa Concordia where 28 people are confirmed dead and four remain missing at the time we are writing this. In each of these cases, denial, emotion and suppression of the truth (an attempt to preserve reputation) played a role in causing deaths. This is still happening today. Many people would claim these bad decisions were human error and stop there. Better training has been suggested as the answer to these kinds of dilemmas, but that is a countermeasure. A real solution demands that we understand the dynamics of how people behave under stress and how authority and ego interact with keeping people safe.

Investigations of the Costa Concordia disaster also show that management was aware of the Captain's penchant for being a daredevil. So, 28 people are confirmed to have lost their lives because it is difficult to say to a charming captain, "You are not suitable to provide a safe environment for our passengers." Be aware that it is easy to say here and in hindsight because we already know he didn't successfully "dodge this bullet".

Any REAL safety solution must address all of the feelings and social dynamics that prevent real people from doing what is right and safe.

In construction, emotion can create a negative consequence resulting in accidents. Whether it is emotional reaction to a boss, a co-worker, a work condition, or the need to perform rework under pressure, emotions can affect the workplace. A reaction to a condition in a worker's personal life can also adversely affect the work environment. Emotion can cause people to lose focus, take shortcuts, use the wrong tool, and go against their better nature, common sense or company safety policies and practices. Awareness of this effect can be instrumental in putting an end to such behavior.

Make note of any ideas that occur to you about the people and situations that you encounter during your jobs. We have made our notes and we will address this issue later in the book. For now, we are getting a better grasp of how this and similar "accidents" inform us regarding safety and risk situations where emotion is involved.

EXAMPLE FOUR: Three-Mile Island, Chernobyl and Fukishima

This example will be different from our other examples, and thus will illustrate a different way to consider safety and risk. We are going to view these disasters through the eyes of a California high school student who attended school before the Three-Mile Island "accident" and subsequent nuclear disasters occurred. Here is his story.

In chemistry class, his teacher assigned a project related to nuclear energy. The project was to find out the weaknesses in nuclear power and then present them to a nuclear power expert in the form of tough questions. The expert was scheduled to visit the class in a few weeks. The assignment was to understand fully the safety and risks associated with nuclear power. She, the teacher, predicted that based on rational science the class would realize that the reactive and irrational fears of a non-scientific public were unfounded. She also claimed that the commercial nuclear power industry had a perfect track record and never had a nuclear accident or a single death. It was an impressive discourse. However, she did emphasize that the class was to take a fully scientific approach and not to short change logic or thoroughness. Based on their preparations, they were to ask very difficult questions of the speaker and to be concerned more about science than his feelings. The teacher was very enthusiastic.

Can you guess how the assignment went? The first row of students in sequence asked one to three predictable questions and the speaker deftly answered every one. A few of the next students asked really tough questions like, "Since the half-life of some of the radioactive components is on the order of fifty thousand years, how can we be assured this does not leak into the environment?" He quickly countered with studies and scientific language about the stability of salt mines and containers that are radiation hard and will never degrade or break. He reiterated repeatedly that the commercial nuclear power industry has never had a major accident. His understanding of the statistics and the science was beyond the class's knowledge and he projected authority, competence and confidence.

Then our student who has provided this story asked his questions based on thorough library research. (No internet in those days!) His first question was, "How safe are the Rancho Seco and the Three-Mile Island plants when considering the unreliability and similarity of the water cooling systems in government-operated reactors which had partial reactor core meltdowns, 8 major accidents costing over 300 million dollars in losses and three fatalities which required lead coffin burials?" Our student followed with, "Isn't it reasonable to expect and even predict that if similar designs have failed in the past, in the future we can expect a similar outcome?" And he finished with, "What is the one hundred and twenty thousand year plan to maintain awareness of where the nuclear waste canisters are stored and how safe are they considering that even the most robust civilizations only have maintained survivable records for about ten thousand years? Some civilizations lose their records in less than five hundred years. Is it okay for the people one hundred, one thousand or ten thousand years from now to happen on the containers and open them so that we can have cheap power today? Or were these issues not part of your safety considerations? And if they were not taken into consideration, how safe can your predictions be? I am only a student with a few weeks available for research and I discovered these weaknesses. Did your experts - who are responsible for the public's safety - gloss over these issues?"

The room, the teacher, and the speaker at first were silenced. Then the speaker blushed bright-red. Our student thought, "Success! An 'A'! I have stumped the expert with clear, logical scientific thinking and have poked a hole in the nuclear power safety plan! I did what I was asked to do and did it quite well! I wonder, what can be done to address these issues?"

Instead of receiving an 'A', what occurred surprised our foolish and innocent albeit clear thinking student. The teacher and the expert pounced, angrily, using emotion and authority in an attempt to intimidate.

"Was this a setup?" the student wondered. "Was the teacher insincere in her request to have us be scientific and ask hard questions?" Our student got stubborn and did not back down. Instead, he vigorously pointed out a stream of additional facts and weaknesses. The class suddenly rallied behind the rebel-student and many young impressionable minds were lost to the teacher's agenda to convince them that the current nuclear power designs were safe. Eventually, it was clear that the issues mentioned would not be addressed. In a tragic and ironic turn during the semester session, just a few months later, the Three-Mile Island Nuclear "accident" occurred due to a failure in the water cooling system. Of course, this was followed by a string of others including Chernobyl and more recently Fukishima. The "accidents" all have some patterns in common. It is also finally clear that the safety formula for nuclear power is far from adequate (zero harm).

Judging from the recent approval to build the first nuclear power plant in several decades, as a society, we are saying the value of the energy is more important than a few lives. We wonder if those in Fukishima, Japan feel the same way. Our hope is that we can shift the collective mindset to one of deeper problem-solving in safety and not just the obvious first and second steps.

From an inventive viewpoint, deeper problem-solving is necessary to accomplish safety in our construction works, but clearly is hardly understood or believed to be possible.

Our intrepid student was Scott, one of the author's of this book. The lesson he revealed is clear. It applies directly to scientific investigation, problem-solving and the safety plans of any company. In pursuit of the truth through factual investigation, sincere questioning and rigorous logical challenge, the biased defense of system weakness will be revealed. When the use of logic and fact creates a storm of emotional response, know that people will resist even the most obvious realities of a situation.

It is typical for those invested in the status quo to create emotional resistance in order to maintain forcibly the current viewpoint, even if it is unsafe. If a high school chemistry teacher will do it, then we might also expect that some construction leaders may also do it. Of course we are using this knowledge to create awareness so perhaps the dynamic in construction safety need not play out this way.

It takes great personal perseverance to proceed and generate real solutions. Most people cannot endure it especially under the assault of a fearful authority. The ones that do persist are a treasure; often penalized rather than rewarded - until they succeed.

To get to the next level of safety (zero accidents - zero harm), one must face reality as it exists, not what one wants to believe. One must face the fact that you have beliefs or perceptions that may or may not describe the whole situation. One does NOT need to be an expert to call attention to pending failure. Clear-thinking, honest research and sincere introspection are often more important than rigid expertise when it comes to improving safety.

EXAMPLE FIVE: Pressure Vessel Explosions Causing Deaths and Injuries [21]

At the previous turn-of-the-century, many deaths occurred as a result of explosions. According to the American Society of Mechanical Engineers (ASME) from the mid 19th century into the early 20th century:

> *"For want of reliably tested materials, secure fittings and proper valves, boilers of every description, on land and at sea, were exploding with terrifying frequency… Engineers could take pride in the growing superiority*

---

[21] John Varrasi, ASME Public Information. "*The True Harnessing of Steam*," March 2011, ASME website, https://www.asme.org/engineering-topics/articles/boilers/the-true-harnessing-of-steam (accessed on 28november 2012 and 5september2013).

Also, Domenic Canonico, Ph.D., Former Chairman, ASME Boiler and Pressure Vessel Committee. "*The History of ASME's Boiler and Pressure Vessel Code*" March 2011, ASME website, https://www.asme.org/engineering-topics/articles/boilers/the-true-harnessing-of-steam (accessed on 28november 2012 and 5september2013).

"*The National Board of Boiler and Pressure Vessel Inspectors Inspection Guide*" Viewed at the following URL: www.nationalboard.org/SiteDocuments/aboutus/inforguide_web.pdf (accessed on 28november 2012 and 5september2013).

And, "*National Board of Boiler and Pressure Vessel Inspectors*", Wikipedia, (accessed 6december 2012), http://en.wikipedia.org/wiki/National_Board_of_Boiler_and_Pressure_Vessel_Inspectors

of American technology but they could not ignore the price of 50,000 dead and two million injured by accidents annually."

The dramatic expansion and use of steam as a power source pushed the economy of the world forward. Meanwhile, several minor and major explosions were killing people and the actual root causes were not understood. The value of having steam was considered more valuable than the occasional lives lost. However, that was not true for everyone.

In 1880, a group of pioneers in technology met voluntarily to discuss and address major concerns in the engineering field. Alexander Lyman Holley, Henry Rossiter Worthington and John Edison Sweet led what was soon to become the American Society of Mechanical Engineers (ASME). The first major issue they chose to address was the safety of pressure vessels. To reiterate, it was a time of rapidly advancing technology and these men had the skills to generate additional millions in personal wealth and they *chose* to spend their time protecting the public.

Thus, the legacy of arguably the greatest volunteer-driven engineering society was begun. As ASME leaders addressed issues, they created a standard that exists today to serve and protect the safety of the public. The Boiler Testing Code in 1884 was created. While this did help reduce accidents — deaths and injuries continued to occur. Unfortunately, in 1905, a dramatic "accident" occurred at the Grover Shoe Factory in Brockton, Massachusetts on March 10, 1905. The explosion of an old boiler leveled the factory causing fifty eight deaths and one hundred and seventeen injuries.

In 1911, ASME formed a new strategy to eliminate deaths and injuries by creating the ASME Boiler and Pressure Vessel Code (BPVC), which focused on all aspects of pressure vessels including design.

This made new boilers safer, but accidents still occurred with old boilers. Therefore, once again people of honor chose to step up and eliminate these safety issues. In 1919, Ohio's Chief Inspector met with chief inspectors from various jurisdictions to develop a consistent reference of construction standards and to create an inspection board with the authority to regulate. Since 2002, North American inspectors checked over six million pieces of pressure equipment resulting in more than 556,000 potential accidents prevented (that's about one violation every ten inspections). The results are impressive for a countermeasure.

Over the years, the number of deaths and injuries fell. ASME remains the leader in setting the standard for pressure vessel safety. Remember, this was largely a volunteer infrastructure and effort to make pressure vessels safer.

So what can we take away from this examination of history's subtext or psychological inertia? The key lessons involved are:

- At every decision point it takes a [choice + action] to improve safety;
- Safety requires continual persistence to learn and keep understanding at deeper and deeper levels;
- When one kind of improvement does not deliver all of the expected safety results, new subject matter knowledge needs to be investigated. Problem-solving must continue to achieve the desired results; and
- It takes leaders to move safety forward, whether they have positional power or not.

Are there still deaths and injuries due to pressure vessel safety problems? The answer is yes, but over the past decade deaths have fallen by over twenty five percent and ASME remains a leader in continually advancing safer programs.

EXAMPLE SIX: "Big Blue" Crane Tragedy [22]

In July of 1999, there was a major accident in Milwaukee, Wisconsin during the construction of a new stadium that involved a large crane known as "Big Blue".

On a windy day a major crane 'pick' needed to take place, i.e., a large fabricated section of the roof structure weighing over 400 tons was scheduled to be hoisted to the top of the building. Word-of-mouth has it that one operator felt there was too much wind and refused to make the 'pick'. Instead of listening to his experience, the supervisor replaced him with another operator. Earlier in the day, a group of ironworkers went home because they felt it was too windy to work... an interesting fact.

On the day of the crane collapse, there were OSHA representatives inspecting fall hazard concerns on-site and they began videotaping this critical 'pick'. The crane

---

[22] This information for this event came from these three sources:

US Department of Labor, OSHA Quick Takes; *Safety and Health Case Study: The Great American Ballpark - OSHA*; (accessed the final time 31august2013) https://www.osha.gov/dcsp/success_ stories/ compliance_assistance/abbott/abbott_casestudies/slide30.html as well as https://www.osha.gov/ dcsp/success_stories/compliance_assistance/abbott/stadium_construction.html

Also, *Lampson Transi-Lift "Big Blue" Crane Collapse*; an analysis by Brian Riewestahl, E.I.T., M.S., Penn State, 2010 (accessed the final time 31august2013); http://failures.wikispaces.com/Big+Blue+Crane+ Collapse

Also, *The New Crane Standard* by Molly O'Brien from FDRsafety, a Workplace Safety Consulting company, December 2012; (accessed the final time 31august2013); http://www.fdrsafety.com/wp-content/uploads/2012/12/FDRsafety-CraneStandard.pdf

operator raised the load off the ground sufficiently so that the rigging configuration could be examined. Then they started the pick. Because of "ground failure", the Lift Superintendent directed Big Blue to move to a secondary location. This location increased the difficulty of the lift. Once moved to the new location, the crane began raising the fabricated structure over the top of the section already in place (this is not what was supposed to happen, but the new location predicated this movement). Suddenly, onlookers heard the squeal of brakes and then a "cracking sound". A voice (on the OSHA video) said, "What the hell was that?"

The "What" was the beginning of a catastrophic crane failure. In the video you can see the crane boom and its load starting to collapse. As it collapsed, the boom fell across the boom of another crane that held three men in an elevated platform. They fell to their deaths that day because of being in the wrong place from a safety standpoint. The tragedy is a risk-laden event which should not have happened.

Insufficient wind monitoring at or near the elevation where the pick needed to land was a big factor for this crane failure. Side winds brought a dynamic force to bear on the load at that height.

Also, the second crane held three men in a man-basket underneath to bolt the section once it was in position. But that second crane was now inside Big Blue's boom radius during the tough pick. A fatal error!

To compound this tragedy, only two ironworkers were supposed to get on the raised platform and at the last minute a third decided to join them. Such a split-second decision added one more victim and one more family to grieve their loss. These were senseless deaths that should never have happened.

This was a high-risk environment in which there was no room for error. On that fateful day, several things happened that altered the original plan. Bad decision-making led to three men losing their lives and nothing can change that. You can ask this question; "Was the Project Safety Program at fault?" Certainly failure to comply with OSHA requirements played a large part in this crane disaster.

After the crane collapse, OSHA lists the following violations that go to root causes of this accident:

- Failure to factor wind into the crane loading
- Lifting workers during high winds
- Three people in the personnel platform (so many people were not required to observe the work being performed)
- Failure to follow the manufacturer's limitations on the crane
- Lifting loads in excess of the crane's rated capacity
- Not keeping workers clear of suspended loads

- Failure to properly calibrate the crane's load indicator
- Improper ground loading conditions (unstable and out of level tolerance)

The regular crane operator of Big Blue used good judgment when he refused to make the pick. We say this because we know that the operator in the seat always has the last word at the moment before any pick is made.

We suspect that production got in the way of safety and clouded good judgment. This is important to note because this contradiction probably often gets in the way of reaping good consequences.

There also could have been personality issues related to pride and a possibility of positional authority or power.

To eliminate such tragedies, we must better understand the psychology of positional authority and the psychology of choosing performance over safety. We need to find a more useful psychology and use that.

We could talk about the devastating toll to family and friends due to loss of life, but let us look at this accident from a purely financial aspect for the bean-counters among us. Approximately $100 million was spent to make the repairs due to the damage done by this crane disaster. In addition, the three widows received around $1.75 million apiece and punitive damages were awarded in civil lawsuits between $27 and $97 million dollars[23]. As well, the project was completed one year late and was way over budget. These were not the consequences a good business person would desire to have on their resume.

This accident is infamous. And yet, not to be insensitive, but we sometimes hear people say, "it was only three deaths" or "it could have been worse".

NOTE: *This highlights a hidden mindset and that is why we are noting it at this time, and in this way.*

Despite it sounding callous, we can use this information to solve deeper levels of problems. For example: "It was only three deaths, not bad when compared to 58 deaths." *(Do you notice how the use of comparison allows for detachment and reduces the pain of responsibility?)* Here's another: Twenty eight people were killed but we got over 4,000 passengers off the Costa Concordia alive." And here is a slightly different version: "We delivered on-time even though our safety record was less than perfect. One guy

---

[23] We suggest that this money would have been better spent as Research and Development dollars to benefit the construction industry so this kind of accident could be prevented and these families would still have their loved ones in their lives.

(notice there is no name attached to the statement, which allows for detachment) lost an arm and another hurt his back, but no lives were lost and we avoided that huge late delivery penalty (liquidated damages)." Are you noticing a few patterns within this hidden mindset? To get to the next level of safety (zero accidents), we must address these hidden patterns head on.

Hindsight often clarifies the situation, but let us be truthful; hindsight was not necessary to make the correct decision during the Big Blue Crane Incident. Every safety manager knows that the man-lift platform should not have been occupied within the boom radius of Big Blue. They knew it at the time, too, but they were focused on the wrong part of their safety vs. profit equation.

Previously we asked a question: "Are there master patterns to how accidents occur?" Can you see the similarities in the tragedies thus far discussed in this chapter? If you see any patterns in these accidents, especially if you also notice patterns between these accidents and your construction projects, make a note of them - at this point, that's all that is necessary.

EXAMPLE SEVEN: City Center Projects, Las Vegas, Nevada

A newly-hired reporter was handed her first big investigation into the troubling patterns of accidents during the 2008 Las Vegas construction boom.[24] There were

---

[24] "*Pace is the new peril: Amid pressure to finish massive projects, 9 men have died in 16 months*" by Alexandra Berzon; March 30, 2008, Las Vegas Sun News. This story was one in over 70 articles and editorials published by journalists at the Las Vegas Sun News. We used a combination of these stories for the book. For excellence in investigation and coverage of construction deaths on the Las Vegas Strip (including lax regulation enforcement and the government, management and labor unions failures to protect workers), the Las Vegas Sun won the Pulitzer Prize for Public Service. This is the journalism industry's most prestigious award. Alexandra Berzon was singled out for in the award for her "courageous reporting" despite intimidation and lack of cooperation. The news agency's coverage helped force federal investigations and instigate congressional hearings, which led to Labor Dept reforms. http://www.lasvegassun.com/news/2008/mar/30/construction-deaths/; and http://www.lasvegassun.com/news/2009/apr/21/sun-wins-pulitzer-prize/; http://www.lasvegassun.com/news/topics/construction-deaths/ (We first accessed this series of articles and websites on 12june2012).

We also used The Journal of Safety Research 41 (2010) 263–281; *[Case Study] CityCenter and Cosmopolitan Construction Projects, Las Vegas, Nevada: Lessons learned from the use of multiple sources and mixed methods in a safety needs assessment*; by Janie L. Gittleman, Elizabeth Haile, and Pete Stafford from CPWR the Center for Construction Research and Training, Paige C. Gardner, Julie M. Sampson, Erica D. Ermann, and Peter Y. Chen from Colorado State University, and Konstantin P. Cigularov from the Illinois Institute of Technology.

And we used the Review of the Nevada Occupational Safety and Health Program October 20, 2009; http://media.lasvegassun.com/media/pdfs/2009/10/20/final_nevada_report.pdf (last accessed on 10/28/2012)

several projects being built within a small area. At the center of the problems seemed to be the megalithic City Center Projects under pressure to open by a specific date. The projects' scope was record-breaking. The pace of work was chaotic. The regulatory agencies were understaffed and it was difficult to find enough safety professionals to oversee the projects. The construction area was congested with people, materials, and equipment - too much happening too fast with no time to do it the right way the first time and certainly no time to rework it safely. Oversight was minimal at best. At one time there were up to 7,000 workers; crews were running 24 hours around the clock. There seemed to be too little time to make sure the crews were properly trained or to perform regularly scheduled maintenance, housekeeping, or to enforce basic safety practices. See examples below.

EXAMPLE ONE: On August 10, 2007, a worker was sent to lubricate the secondary manlift (elevator) workings. He was killed when the descending counterweight from the other elevator came down on him. OSHA determined that workers had not been trained to lock out the lift system before attempting to service it. Also, operators and supervisors were not following manufacturer's guidelines for maintenance. Six violations added up to a $21,000 fine for the prime contractor, but during an informal, closed-door meeting with the contractor, OSHA withdrew all citations for a total fine of $0.

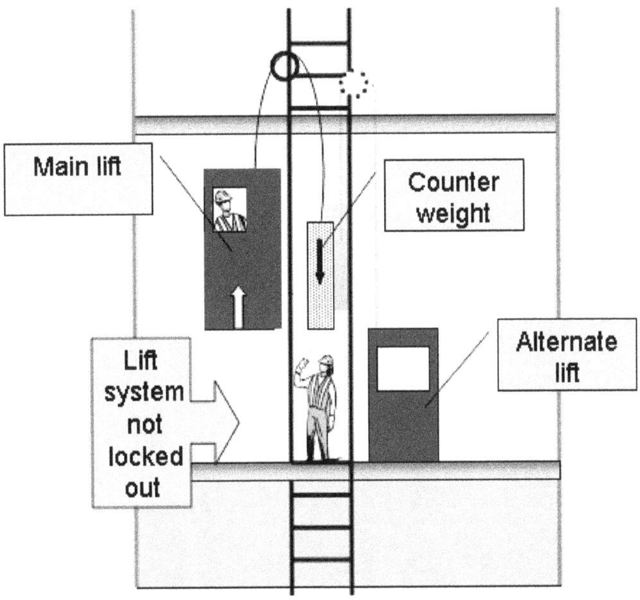

This drawing is based on the Las Vegas Sun News' concept drawing

Total Fines for This Fatality Accident Was $0

EXAMPLE TWO: A crew made a large hole in some metal decking then left with the hole uncovered. A worker tragically fell through the hole, over 50 feet, and died. It may have been avoided if he had his safety harness attached to a railing. OSHA found the hole should have been properly covered or should have been barricaded with warning tape. Also, there was no safety netting below the hole in case someone fell. There was also no attachment point to tie off to with a safety harness and no personal fall arrest system available. The worker's PPE was faulty; it did not have hooks on the ends of the harness so he could not have made the attachment. OSHA found 3 violations and fined the contractor $13,500, but in an informal, closed-door meeting with the contractor, OSHA withdrew all citations for a total fine of $0.

EXAMPLE THREE: Four men were helping to remove a crane from the support forms for pouring concrete into the elevator core. Two workers were on top of the forms and two were inside the elevator core beneath the forms. The forms weighed 7,300 pounds. When the crane was unhooked, the walls buckled into the elevator core killing two of the workers. OSHA came to the site to inspect the conditions surrounding the accident and found that the crew had not been trained properly on the installation of forms as there were no bolts, pipe bracing, or other required attachments to hold the forms securely in place. Also, the inspector found that the workers had not been properly trained to remove the supports. OSHA found that insufficient attachment points could not support the walls once the crane was released so the forms buckled inward killing two of the workers. The prime contractor was fined $14,000 at the time. However, in an informal follow-up meeting between OSHA administrators and the prime contractor, OSHA withdrew half of the fines, which is not against Nevada law but was definitely against OSHA general practices. Observers must ask why this and other accidents happened in the first place and why OSHA was so unwilling to impose proper deterrents.

Form Fell Inward; Two Workers Were Killed In This Accident

EXAMPLE FOUR: A worker, wearing the correct PPE was correctly tied off to a beam as there was no other place to tie-off on the surrounding structure. He was "bolting-up" a spreader I-beam inside the elevator core on the fourth level. He was correctly attached to the I-beam when it broke free off the wall plate and fell 44 feet. He died. OSHA found 6 violations and fined the contractor who installed the beams $12,150. Although OSHA could not absolutely determine how many bolts had actually held the steel in place, the evidence indicated that only a single ¾-inch bolt held up each end of the beam. There were no nets or scaffolding platforms in place, which was required by safety regulations to have been installed. This appeared to be either substandard work or unsafe working conditions.

These examples cause observers to wonder at the seemingly callous disregard for human life. The mindset of "that's the cost of doing business; it's a dangerous job" must shock most observers. It seems that an experienced prime contractor - even if they only had an average safety record - would not let the situation get so out-of-hand. It begs an answer: how could these large companies allow these types of repeated violations; and how could the unions seem to fail in their workers advocacy tasks; and how could the state regulatory oversight agency allow so many blatant and repeated violations? It is rather unbelievable for all this to happen on a construction site today.

EXAMPLE FIVE: But let's be even-handed in our observations. Many of the craft employees who worked on the City Center Projects contributed to an unsafe working environment, too. Workers spoke of exhaustion from massive overtime but continued to volunteer for more because the money was so good. Drivers fell asleep while transporting materials. People knew they were not working with full attention. It was also reported that numerous workers had drinks at nearby bars during lunch, before work or between shifts. This was a clear violation of safety protocol and something that every worker should realize is neither in their best interests nor the best interests of the people they work with. There were not enough safety managers to monitor this problem although it was repeatedly reported so it continued unchecked with nobody taking enforcement responsibility.

To make matters worse, industry observers said that both the unions and Nevada State OSHA were seriously lacking in the will to pursue proper remedies. Inspectors told federal investigators that they were discouraged by management and legal council to find a "willful" violation, which carry high penalties. Only one willful violation was made during the period investigated, but it was later reclassified to a lower category. In each of the deaths attributed to this project, OSHA discovered many violations, but instead of forcefully upholding safety laws, state OSHA minimized the citations or withdrew them completely after meeting with the contractors - usually in private meetings without the victim's family or union representatives present.

Safety experts around the country stated that the trend of reversing findings and withdrawing citations is disturbing - especially in fatality cases. OSHA enforcement policy exists to deter serious and repeat violators of the law through stiff punishments. But in Nevada OSHA regulations, the burden of proof appeared to be completely backwards. Usually it's the employer's burden to provide proof. Typically, if the employer contests the official finding on a serious violation, OSHA might reopen the case, send it to the review board, or let the courts make a ruling. Most OSHA administrators would not back away in fatality investigations because withdrawing findings and removing citations on dangerous projects sends a strong, clear, wrong message - that it is of no consequence to perform exactly the same in the future.

Is it unreasonable to ask, what kind of mindset was in place? What kind of trade-offs and assumptions were considered acceptable throughout this project and at critical points along its completion? There was so much money and reputation and momentum riding on the status quo mindset (which was: get 'er done as fast as possible with as little oversight as can be managed because that is best for everyone).

Standing up against that mindset must have seemed overwhelming to the intrepid reporter for the Las Vegas Sun and other employees, managers, and family members watching the abuses but feeling powerless. It takes extraordinary integrity to stand up for a safe environment when those in power act as if the major priority is production. Just imagining what the evidence below "feels like" conveys the backwards value the whole city seemed to place on safety.

- No Fines or reduced fines, even when it meant going against OSHA policies.
- Extreme overtime work; people were falling asleep on the job but still signed up for more.
- Stories of people drinking on the job; it was reported, but not stopped.
- People getting killed because the pace of work was so frenzied; there was no time for safety and not enough oversight.
- Closed door meetings are conducted without the bereaved family's input; the union representatives also did not attend.
- No apparent consequences; it looked like everyone was running free; they could do anything they wanted. It looked like the law didn't care, the unions didn't care, the other workers didn't care, the city didn't care….
- Push, push, push on production
- Safety protocols were routinely ignored.

Do you feel it?

Upon facing this issue, we turned to our structured innovation toolset to find others who have stood up against great odds and a pervasive and dark mindset. We found some examples.

Eliminating the "production comes first and safety second" pattern may be just as difficult as the challenges faced by the Civil Rights Movement or Women's Suffrage who labored to change the mindset of a society. Therefore, these movements may provide some useful hints of how to conquer stubborn attachment to perverse principles.

This is an important idea and worthy of emphasis. What is needed today is a change in thinking about safety. It needs a radical overhaul just like the America's Civil Rights movement shifted institutionalized racism or the Women's Suffrage movement gave unalienable freedoms and citizenship duties and rights to over half of the American population.

There existed, before these movements, entrenched wrong-thinking so pervasive the priorities were not seriously questioned - very similar to the entrenched mindsets of gambling on safety during the City Center Projects in Las Vegas. This reminds us of the ancient beliefs in construction, when slaves performed the work and their lives were considered expendable.

It appears that a similar stubborn elitism that denied freedoms to people of color and to women also denied the guarantee of safety on these construction sites. It was and still is difficult to shake out these wrong-headed belief systems that we as a society or as an industry are conditioned to believe. The worst part is that it appears that everyone around us is in agreement due to their actions or inactions.

It is not so hard to understand that when the pace is frenzied and the momentum is full-steam ahead and the excitement of a lot of money to be made is held out like a tantalizing tease, then we can see how the City Center Projects went too far out on the safety cantilever.

NOTE: Structured Innovation has a rule of thumb stating that 80% of the problems anyone faces have been solved in another industry or field of study. These movements have much to teach about changing people's minds and hearts. Also, notice your reaction to comparing construction safety to the Civil Rights Movement or Women's Suffrage. Do you find yourself feeling defensive or seeing opportunity? Note the nature of your reaction.

Consider how the leaders of the Civil Rights or Women's Suffrage movements were willing to stand up to harsh criticism, go to jail and sacrifice everything to prevent injustice from prevailing. They didn't back down. And issues of social injustice will come up again and again in society as we grow our consciousness and refine our understanding of what matters most. So we need real answers.

Later in the book we will discuss methods and ideas that will help reveal why injustice and harm occur; and how to open your mind to see what is hidden. We will show you how to articulate all of what you want and provide a strategy to get it: a safe environment and a profitable company.

Doesn't it make more sense to solve the actual contradiction of safety versus production than to have an industry-wide fight filled with anger, denial, suffering and remorse while leaving money on the table? Perhaps you can see why we are passionate about this profound idea of resolving contradictions that are fundamental in human priorities.

Just like the success of other movements depended in large part upon the commitment of leaders, so does the success of workplace safety depend upon the commitment of construction 'leaders'. That means every single person must be a recognized leader for their own and others' safety. Every single person must act to solve these stuck unsafe thinking patterns. You must identify contradictions. You must solve tough problems. To do anything less is an admission that human life is expendable.

Remember that newly-hired staff reporter at the Las Vegas Sun News, Alexandra Berzon. When her Deputy Managing Editor, Drex Heikes, asked her to look into the "unusual number of construction deaths on the Strip" she took the lead on the story. She refused to give in to government bureaucracy, threats of harm, cover-ups and that pervasive driving momentum - she took it all on because she knew that 'gambling on worker's lives for a few hours progress' was the wrong point-of-view and she was "pugnacious" enough to "not take 'no' for an answer".

Most visitors to Las Vegas will be oblivious to the pain, suffering and tragedy that precipitated the completion of the City Center projects. The many contributing issues that led to deaths during construction make this project, in the authors' view, a disturbing industry-wide safety failure. The mindset displayed by so many people on the City Center Project is a harsh testimony to the power of history, hurry, and psychological inertia.

The devaluation of human life has gone on for far too long with the belief that nothing can be done about it. Further, this mindset remains hidden even in some self-aware leaders, supervisors and craftspeople who by their actions reveal foolish subconscious priorities with unthinking automatic responses. If we are to really prevail in getting to the next level of safety performance (zero accidents), we must thoroughly address all components of a leader's mindset both public and hidden and resolve the deep issues we face.

We are delighted to say, we have powerful answers for you.

EXAMPLE EIGHT: The Golden Gate Bridge's Safety Net

Unlike other huge construction projects built in the 1930's (such as Hoover Dam and the Hawks Nest hydroelectric diversion tunnel built near Gauley Bridge, West

Virginia) the men charged with the task of building the Golden Gate Bridge were pioneers of safety. [25] Chief Engineer Joseph Strauss was determined to create the safest possible working conditions for his crews despite the high winds, freezing temperatures, blinding glare, treacherous fog and rain, and dangerous heights. He researched and ordered up-to-date safety innovations at significant cost while other companies were refusing to furnish even the most basic personal protection gear.

Strauss required that every worker wear a "Hard Boiled Hat" made of seamed canvas and glue or tar and hardened and then painted black - the Bullard hat. Falling objects might hit a man wearing that hat with force enough to break his jaw, but it would leave the worker's skull undamaged. This was progress indeed.

He order special goggles that helped defeat glare, special creams to combat the endless winds, and respirator masks so riveters didn't breathe in lead fumes as they put in approximately 1.2 million hot rivets through lead-based paint. Not only did he provide the personal protection gear, he demanded workers wear the gear or leave the job.

Strauss also demanded all workers wear safety lines and that his workers adhere to a strict safety code - no daredevil stunts, no working with hangovers, and no diving into the safety net for thrills.

In 1936 he tried a novel safety device - a $130,000 safety net suspended under the bridge from pylon to pylon. The manila, square mesh net extended ten feet wider than the side trusses and fifteen feet further than the roadway's length. The net saved 19 men from falling to their deaths (these men called themselves members of the "Halfway-to-Hell Club").

In the days prior to Strauss' safety practices during the construction of the Golden Gate Bridge, the typical rule of thumb was one fatality accident expected per million dollars total cost of the project. Imagine using that rationale to predetermine how many workers would die during the construction of your project today and not also generating a solution to prevent it.

---

[25] We read a number of accounts of this story but these are the two main sources we are citing:

" '*Cheating Death': Worker Safety During Construction*"; essays (author unknown) from the research library of the Golden Gate Bridge Highway and Transportation District website: http://goldengatebridge.org/research/CheatingDeath.php (first accessed 06december2012)

We also used the *PBS' American Experience: The Golden Gate Bridge* broadcast and article at the website: http://www.pbs.org/wgbh/americanexperience/features/general-article/goldengate-safety/ (first saw the website in 2012, last accessed on 31august2013)

The cost to build the Golden Gate Bridge was between $30-35 million. Construction took place between the years of 1933 and 1937. For three years the safety record on the project was dramatically beating the number of fatality accident odds - without any fatality at all until October 21, 1936 when Kermit Moore fell to his death. Despite Strauss's innovative best practices and rigorous safety codes, tragedy again struck the project on February 17, 1937.

Eleven men were working on a five-ton wooden platform near the north tower. Two more men scraped debris in the net below the scaffolding. One side of the platform gave way. Worker, Tom Casey, managed to grab hold of a beam and he hung there until he was rescued.

For a few moments, the platform dangled from the bridge then fell into the net along with the rest of the men. The net held, but only for a few more moments before the weight tore a hole in the manila webbing. Twelve men fell into the water "twenty-two stories below". Two men survived the fall. Ten men did not.[26] Sadly, tomorrow never came for those ten workers.

The subject of the Golden Gate Bridge safety program is brought up for three reasons. The first reason is to honor the dedicated efforts of Chief Engineer Joseph Strauss and his team. When other projects were hiding root causes of accidents for insurance benefits and focused on squeezing out more time and money at the expense of worker health and safety, this team of amazing men walked their talk, invested in safety innovations, held everyone accountable, and brought in a world-class project under budget, better than on-time, and much safer than other projects from that era.

The second point is: even when you are sincerely acting toward making safety a top priority and are using the best possible methods old and newly invented, we still may not know enough to keep our crews safe. Could Straus have anticipated that a five-ton platform could have fallen and broken through the net? This is a painful question because any safety professional pushing the state of knowledge must face the challenge of figuring out what they do not yet know. This is why it is important to think inventively when it comes to safety. If you think inventively, you can invent failures that have not yet happened. Therefore, you know it is possible and then can prevent it. This is why we are pushing so hard to make inventive thinking an integral part of safety.

---

[26] O.A. Anderson, Chris Anderson, William Bass, Orrill Desper, Fred Dümmatzen, Terence Hallinan, Eldridge Hillen, Charles Lindros, Jack Norman, and Louis Russell are the ten men who did not survive that day.

The third reason we mention the Golden Gate Bridge project is because there needs to be a 'safety net' in place on every construction project. This is being used as a metaphor to make a point and hopefully helps provide you with imagery that you can use.

To actively support safety processes in the workplace means that you are a part of the safety net that must be in place for every worker. You need to be visibly involved as a safety example. You need to keep an eye out for new innovations that work to keep crews safe. You need to set aside your ego and adopt other people's suggestions when they are better than yours, like Strauss did. You must speak up, even in the face of intimidation and frenzied schedules.

It behooves us to remember those who have lost their lives in the course of doing construction work. Life may go on for most but the family and friends of those who are killed in the workplace continue to experience pain and suffering. Their lives are irrevocably changed and often some never know "normalcy" again. In and of itself, this ought to be motivation enough for everyone in the construction industry to strive constantly and be on task seeking safety perfection wherever they work. Again, there needs to be a tomorrow for everyone working today.

Bottom line #4: "Accidents" occur near hidden mindsets, denial, high emotions, indifference, compromise or suppression of the truth.

Safety Mantra #4: ***Negative emotion and safety are like oil and water; they don't mix.***

Take-Away Points: ***Chapter 4***

- Any REAL safety solution must address all of the feelings and social dynamics that prevent real people from doing what is right and safe.
- Within "accidents" are master patterns that reveal the mechanism of how failures occur.
- In the past there was a mindset that people are expendable. Sadly, some of this attitude prevails today as a hidden mindset.
- History and precedent helps the unaware justify behaving without thought regarding safety. This psychological inertia must be overcome.
- Accidents can affect hundreds of people - not just the injured parties. Yet, if we distance ourselves from this information and set priorities based on that distancing, we will create less safe worksites.
- Fear of authority, misperception of risk, strong emotions, and denial adversely affects people's reasoning and are dangerous indicators that safety is already grossly compromised.

- Inventive thinking helps you invent failures that have not yet happened. Therefore, you know those failures are possible and then can prevent it. This is why inventive thinking needs to be an integral part of safety.
- Advancing the art of safety into science will require deeply understanding how to transform the "mindset of production over people" into something more useful, whether it is hidden or overtly expressed.

"Keep your thoughts positive because your thoughts become your words. Keep your words positive because your words become your behavior. Keep your behavior positive because your behavior becomes your habits. Keep your habits positive because your habits become your values. Keep your values positive because your values become your destiny."

- Mahatma Gandhi

"Weakness of attitude becomes weakness of character."

- Winston Churchill

Chapter 5 - Current Best Practices of Attitudes

On his projects, one of the authors (Brion K. Hanks) often relates this safety goal and commitment "Nobody gets hurt… and safety begins with me."

Let's get clear on this fact: safety is everyone's business. It permeates all aspects of our lives. Whether we are at work or play, and whether we are consciously aware or acting unconsciously, we want our family and our co-workers and our friends and ourselves to be safe.

This chapter details some of the attitudes that the safest workers have on construction sites today. You must develop and demonstrate all of these attitudes consistently every day on the job.

- Attitude of Responsibility
- Attitude of Leadership
- Attitude of Commitment
- Attitude of 'The Student'
- Attitude of Sincerity - No Rebels, No Royalty
- Attitude of Following the Rules with Flexibility
- Attitude of Teamwork
- Attitude of Communication
- Attitude of Follow-Through
- Attitude of Quality

Attitude of Responsibility

We would hope that you can agree with the following statement: "Safety is my business because I care and I'm passionate about putting all aspects of safety first in everything that I do, both in my personal and professional life." This is a good primary mission statement for everyone. Safety First and Safety Always is a good motto to promote: Safety First (because it is the first priority) and Safety Always (because it is always on your mind and always where you take action from).

Through individual integrity we should live it, breathe it and by dogged determination commit ourselves to accepting nothing less than always putting safety first in everything we do.

Once again, if a little voice inside you said, "Yes, but..." make a note of what resistances you automatically had. Later in the book, we will give you information that may help you deal effectively with these hidden triggers.

Employers and workers alike have only paid lip service to safety in the workplace for far too long. Too many workers do 'the minimum' when it comes to safety and the same is true for employers. That indecent notion is a mistake. It is contributory to people getting hurt or killed in work environments. This is a flawed notion and must be discontinued.

Is it true that a construction company can only survive if safety concerns are put second to production schedule concerns?

Make note of your automatic response.

No one wants to come right out and say that this is true, but it sure seems to course through the veins of the industry. The continuation of such a view borders on the irrational and it is why the safety culture needs fixing.

Not only is that compromising belief unseemly, not only is it inappropriate, it is flat out wrong for employers to think that they can expect to prevent accidents in the workplace by continuing to operate as if they can do nothing more. And we can say the same for workers and supervisors. It is inappropriate to make decisions and take actions that cause accidents. Such flawed thinking is an indicator of poor and frankly foolish management and worker choices. It must change.

From the top down (and bottom up) there has to be unwavering leadership that holds back nothing with regard to safety. Commitment means that management, frontline supervisors, and workers will put their reputations on the line and spend the necessary dollars and take the necessary time in order to show that they fully accept the responsibility for worker safety.

In summary, it comes down to this: do you (management, frontline supervisor, team leader, and daily worker) care a little or do you care a lot about worker safety - and your own safety and the safety of your team? This is a question that each responsible person must answer for themselves.

To care a lot means that you will individually contribute unrelenting and obvious positive action to promote safety in the workplace. You will strongly hold yourself and other people accountable. Any degree of doubt, delay or dereliction on your part means that you are part of the problem.

Right human relations demand it and anything less than 100% success means that management and each individual have not done enough. Is that a moral failure? Does poor leadership send a mixed message to the people working in the field, to your peers, and up the ladder? Do people really care? These are fair questions that merit consideration.

Be Your "Sister's and Brother's Keeper" [27]

A subset of an attitude of responsibility is to "Be your Sister's and Brother's Keeper". This is the most surprising and effective countermeasure in a culture of safety today. What does this mean? In a nutshell it means you accept responsibility for yourself and for others.

> *"If we sit by and become complacent and put our heads in the sand, we're complicit." - Shelley Morrison*

EXAMPLE ONE: Brion K. Hanks, one of the authors of this book, was the oldest of five children. If you are the oldest sibling, I am sure that you have heard, "Watch out for your brothers and sisters." What he now shares was his greatest safety failure. When he was nine years old, there was a function going on across the street in a neighbor's yard. Brion and his brothers and sisters started across the street. His sister Deborah, the youngest at three years of age, was naturally the slowest and had started late to cross the street. Like a dog who gets focused on something, all Deborah saw was everyone in the neighbor's yard. She didn't see that a car was coming. The driver apparently didn't see her. His family's good fortune that day was that Deborah did not get run over by the car, rather, she literally ran into the side of the car and was knocked down. She suffered scrapes and bruises and was obviously shaken. It shook-

[27] The original source of the term "my brother's keeper" stems from the ancient biblical story of Cain and Abel. Throughout the ages it has served to remind us of our responsibility to care for each other regardless of our religious persuasion.

up Brion's whole family, but most of all, it deeply affected him because he had failed to watch out for his brothers and sisters.

Brion's parents launched their own root cause investigation in order to learn what happened. Then, together, they developed a plan to prevent such an accident from happening again. Brion's parents made it a family requirement for all siblings to walk to the corner (no matter how far away and inconvenient it might be) and look both ways before crossing at the intersection.

Going forward, when Brion heard, "Watch out for your brothers and sisters," he took it very seriously. With this trait strongly a part of his safety ethic, Brion continued to "watch out for his brothers and sisters" upon achieving greater responsibility during his construction career. This was a guiding first principle, and strong core value for him as he became an engineer, then foreman, then superintendent and ultimately a dedicated Project Safety Manager.

The danger of "knowing better but doing nothing" is that a worker may get seriously injured or killed on the job. An unsafe condition is either corrected or someone is told about it so that it gets resolved right away and is not left as a trap for someone to get hurt.

> *"As soon as you see a mistake and don't fix it, it becomes your mistake."* - *Author Unknown*

Companies must develop a culture where caring enough to correct an unsafe act or condition is valued and admired - even when the "lowliest worker" corrects the "mightiest manager". Is this something that could happen on your projects? Is it a reality, not just in words, but in the hearts and minds of all employees?

EXAMPLE TWO: One day Brion noticed a worker committing a minor safety infraction and a co-worker observing it but doing nothing to correct it. Before Brion could get to the two men, the workers were on to their next task. Brion proceeded to deal with the unsafe worker and then he had a conversation with the observing worker. "I'm displeased," Brion conveyed, "with the fact that you did not correct your co-worker. You let the matter continue and did nothing to resolve the unsafe act. Why didn't you say something?"

The worker stammered and finally said, "Welllll, *you're* the Safety Manager!" Brion gave a verbal warning to both workers and reminded them about Kiewit's First Rule of Safety.

That was on a Friday. Over the weekend Brion thought about that comment and came back on Monday where he spoke to approximately 40 craft workers and conveyed the following, which is useful to any safety professional, "I may have the title of Project Safety Manager but each and every one of you is the Safety Manager over your own safety. You have the responsibility to be observant as you work, do the right thing in order to work safely yourself and care enough to correct a fellow worker when you see them working in an unsafe manner." This became the seed of a new culture of safety - I am my brother's and sister's keeper.

In evaluating the best leaders in the industry, they understand that the collective knowledge and experience of the team is greater than an individual. By being watchful and vigilant 100% of the time, we can help people who are making a safety error. It is a very effective countermeasure that has driven powerful safety results. We have to be our brother's and sister's keeper to perform like the leaders. We must stand for what is logical and effective, even or perhaps especially, when it is inconvenient. Why? Because life and limb are on the line and everyone needs each other to be safe.

EXAMPLE THREE: Brion worked on a project where new employees (less than 30 days on the job) were required to wear a red hat so everyone knew to keep a "safety eye" on that person. Red Hat employees might not know all of the rules. After thirty days on the job, Red Hat employees were reviewed by the Project Manager in order to assess their knowledge and safety buy-in. If they had been actively safe, they graduated to a white hard hat. All persons new to the project went through this, including the safety manager, supervisors, etc. There were no exceptions. It was vital to lead by example. Employees may not have liked the red hat policy at first, but they all appreciated graduation to the white hat.

One day Brion noticed a Red Hat employee crossing a busy section of a worksite road in an undesignated area rather than using the mandated crosswalk area. A loaded forklift was moving along the road at the same time. Thankfully, the operator saw the Red Hat and slowed until the new employee was across the road, and then the forklift continued on his way. Ahhh, but there was a problem.

Brion approached the Red Hat about crossing the road at other than the designated locations where signs are posted. Because the new employee made excuses instead of admitting responsibility, the Red Hat was written up. But that was not the end of the incident. Follow-through and consistency are the hallmarks of a safe environment and Brion is famous for walking every step. He kept an eye on the Red Hat for a good part of the project to make sure the employee learned the lesson.

Brion also radioed the forklift driver and asked him why he had not stopped the worker who was crossing the road at an unauthorized location? The forklift operator admitted his lapse of judgment and said, "My mind was on where I was going with this load." This employee admitted he knew better and admitted he was distracted. Brion simply cautioned him to be more careful and the forklift driver made a commitment to do so.

The forklift driver's supervisor overheard Brion on the radio and later told him, "Thank you for dealing with that. I will speak to all my operators and remind them to be on the lookout for this kind of thing and to deal with it like we require, immediately."

The forklift operator took responsibility for his failure to correct the situation. This kind of consistent and persistent dedication builds a strong safety program on a project and also makes the point about how everyone must do their part in order to keep people safe.

The fortunate thing in this instance is that nobody was hurt.

Attitude of Leadership

It is a current best practice to lead by example, especially regarding safety on the job.

You must realize that people are watching you and modeling your behavior. What you speak is important, to be sure; but what you actually do day-in and day-out is even more important. It is especially significant when you think no one is looking because on a construction site it is likely that someone will observe you. You may never know the ways you influence others; they may never speak of it. But know that your actions have an effect.

Almost everyone believes that safety success is best delivered by way of top-down management's commitment to safety first and foremost; but it is every person's obligation. That commitment must be principled, resolute, and real. Top-down management can demonstrate safety commitment through company communication, but it is the boots on the ground that makes or breaks a safety program.

Once you realize that you are leading by example, it is your responsibility to be very clear about company policies and company safety expectations. Have them spelled out to you in sufficient detail as to leave no doubt about the dos and don'ts in the workplace. After that, you must be consistently accountable for safety. On this there is no compromise.

For example, safety messages can be revealed through policies, on-site signs, and great speeches at company meetings, but it is the in-field follow-through that reduces incidents.

If you are a frontline supervisor, it is up to you to repeat these communications like beating on a drum. And it is up to the workers to take these communications to heart and follow-through with them.

In addition to the example you set, the speed of your response to safe or unsafe actions tells others how serious you are about safety. The strength of the response forms an impression on co-workers and crews. Day-to-day onsite attitudes determine if workers believe in the safety talk that is presented or if they feel it is just lip service.

Talk must be made real. It must be totally genuine to the point that there is no doubt about worker safety being put first and foremost.

It is important to quickly resolve any unsafe condition. Dangerous situations cannot be left unresolved. Rather, they must be dealt with immediately so they are not left as traps to injure or kill someone. The speed of your response 'speaks' loudly.

A wise man said: "You can lead a horse to water but you can't make him drink." When it comes to safety, we firmly believe that each individual makes the choice every moment of every day. Supervisors "bring workers to the trough of safety" through training, setting clear expectations, and helping crews understand that we are accountable, 100% of the time. This is how we build up the belief system. But it is each person's choice. Each of us needs to be self-motivated because we have the power over our own safety and the safety of others.

And here is one of the most important attitudes to have on a construction project. Do the right thing, even when no one is looking. To do the right thing, even when no one is looking, speaks well about your individual integrity. When it permeates to the core of a work team, a company or even a family, then it is a magnet for attracting like-minded people. And, it improves onsite safety.

Attitude of Commitment

A safety keystone is an unwavering commitment to safety success such that people will marvel at that success. Apathy can not rule in the worksite. Ego cannot rule on the worksite. The unwavering rule is every worker must go home safe every day.

It takes total commitment and resolute will to accept nothing less than this. Such a safety culture can be a shining example of phenomenal safety success. It must not be allowed to be rare, rather it has to be real.The greatest thing to remember is that each of us has the ability to make it happen. And why wouldn't we?

Make it personal to know why you are working safe. Take that extra few seconds to work the correct way or to formulate your plan. Is there a loved one waiting at home? Remind yourself of your reasons each time you want to take an unsafe 'shortcut'.

Get to know the people on your crew and your co-workers. What would a safety incident do to them, their loved ones and their plans? Know their motivations; know their situation; know their skill levels; know what they will do before they do it.

Supreme safety success is delivered through the power of choice. Elite athletes choose to continue to do the right thing even in the face of adversity. They do not give up with their first failure or even multiple failures. The allure to be the best of the best drives them to the high level of success. They do not let fear and pain compromises their efforts.

What these elite athletes learn is expressed in the Latin proverb, "vincit qui se vincit - He conquers who conquers himself." Their battle is within themselves. Now you might ask what does that have to do with safety in the workplace? Put simply, such success stems from absolute commitment, drive and the will to succeed always in pursuit of perfection. The same is true for workplace safety.

How badly do you want it? To succeed you must become a master. Safety mastery is possible; however, it takes the will and drive similar to an elite athlete's. And even then, you must keep that skill honed in order to stay on top. The effort of all workers must be voluntary, consistent and motivated by a will that does not allow any safety compromise. Again, what is safety mastery and the constant goal? The Answer: Everyone goes home safe each and every day - safety perfection - because you thoroughly understand safety.

Safety needs to be personal at all levels; it needs to be planned; and full participation must be strongly encouraged by every leader as part of showing genuine concern for the welfare of all workers. Once safety is planned into operations, the workers must understand their part in the plan and act on it.

If you are not decisive and totally onboard with the belief that *accidents are preventable,* then we suggest that you get out of any business that uses the labor of people. If your motives are not pure regarding safety first, then you will be complicit in an accident-prone workplace.

Every person is responsible for their part in working safely to produce good results. Make that commitment. Extreme safety success should never be about "me", it must be about "we".

Demonstrate Commitment

Making the commitment is important as we just said. The next step is to demonstrate that commitment. Make it real. Show it by every action on the job.

> *"Be the change you want to see in the world." - Mahatma Gandhi*

In-person demonstration is very powerful. When other workers see you demonstrating safety and productive work habits, it sets the standards high. What workers see day-in and day-out sets the tone of how 'a good worker acts'. People are watching. Do the right thing always.

It makes a strong negative impact when management holds themselves above 'the safety law'. For example, it would set a horrible example for an engineer or executive to drive an unauthorized vehicle onto the site and park it (just for a moment) in an unsafe location (e.g., behind a crane). And if the bigwig jumped out of the car (without a hardhat or other personal protective gear) in order to 'have a quick word' with the supervisor about the schedule, the workers would have a solid demonstration that safety is lip service only.

The same is true for a respected and experienced worker. If people look to you for leadership and see you working without safety goggles or climbing above 6 feet without being tied off or... then the less experienced workers would have a solid demonstration that safety is being paid lip service only. Observed compromised safety tells the true priority, which is the company's safety program is a sham. Such hypocrisy is what gets people killed or injured in work environments.

However, when workers see their leaders-at-every-level commit to making the workplace safer, it strengthens everyone's commitment. This is necessary and vital to the success of any safety program.

Two of the main reasons leaders pay lip service to safety is (1) they do not believe that accidents are preventable and (2) they believe that safety can get in the way of productivity. Fear causes us to compromise - I want to be productive in order to keep my job. In these cases we follow the rules when we can and when we can't, we cross our fingers and hope we get lucky and avoid getting hurt. This type of thinking actually compromises safety and if you are an in-the-field leader displaying these attitudes, then you are building momentum to compromise safety. If you display these attitudes, you are the problem.

It is up to you to stop this momentum. Leaders at every level of authority must be committed to safety with a dramatic passion that needs to be so infectious that everyone gets it! That means you, whatever your job is, at whatever authority level you are granted. You are responsible and need to demonstrate it!

Attitude of 'The Student'

Each of us is a teacher and a student too. Those with years of experience must be willing to coach and mentor less experienced workers. But less experienced workers have something to teach, too.

We respect new workers because they see with new eyes and have a fresh attitude. Sometimes a new point-of-view and a more productive attitude can overcome resigned experience and entrenched negative situations. Recognizing each person for what they have to offer and for their efforts will deliver increased safety results.

The safe person will always be willing to be a safety apprentice and learn something new that will serve workplace safety. How we conduct ourselves through acquired knowledge, actual field experience, on-the-job training, right attitude and good will is where the rubber meets the road regarding safety success. The challenge is each person's and you are charged with a great responsibility regarding right action.

Having said that, sometimes workers need to learn new skills. Supervisors must understand each of their employees' level of experience. In-field management must also know if their crews understand the company's rules and policies. In addition, supervisors must determine each worker's attitude about that knowledge. Are they cynical or sincere? Safety orientations help install the correct mindset and the workers who resist the safety mindset can be weeded out before they cause injury.

As well, it is important to convey that an employee should never be afraid to ask questions. If there is ever any type of doubt, questions must be asked in order to have clear understanding and confirm that everyone is on the same page and will work to get the job done safely.

Attitude of Sincerity - No Rebels, No Royalty

When the goal is "Nobody Gets Hurt", there is no room for rebels or royalty in the safety world. Some people think that rules are stupid - Rebels. And others think that rules don't apply to them - Royalty.

Nothing is further from the truth. When it comes to safety rules, no one is set apart or above the rules. Gravity, electrical power, and falling objects don't care about your job title or negative attitude. The safety elite are those people who assume responsibility for their own safety and for their peers and are humble and driven in pursuit of perfect safety outcomes - these are noble qualities.

Safety success comes through believers and achievers. Anyone that is not on board must be coached and made to understand that a safe work environment is everyone's responsibility.

Safety rebels make snide side comments and feel like following the rules are stupid or silly. Safety naysayers and pessimists must convert or be cut out like cancer. Anyone contributing to clouds of doubt on this vital purpose cannot be allowed to work around people doing their part for workplace safety. This is the degree of dedication current leaders must have to safety goals.

However, in the future construction site there is a place for those that are skeptical. A proper skeptic is not the same as a safety rebel. They do not look down on safety initiatives, but they do speak up with questions. These skeptics are valuable employees provided that sincere participation and engagement are present along with specific negativity about what won't work. We will discuss "precise negativity" later in the book because this is a very important topic.

Attitude of Safety Teamwork

Unwavering action requires daily effort on the part of everyone in the workplace. Power in numbers becomes more powerful when all are aligned with the same goal. By now you should know what that goal is - every worker goes home safe every day.

At a mass safety meeting Brion said, "Regarding safety on this project, I want us to use the Three Musketeers' motto, 'All for one and one for all.' Are you with me?"

It is important that workers constantly care about fellow co-workers and are motivated to focus on watching-out for each other as their part of workplace safety. When this happens, the project is well on its way to accomplishing the all-important goal of "Nobody gets hurt". Action, commitment, and individual accountability must be expected day-in and day-out, nay, required for real safety success in the workplace.

Workers expect to be told what to do, how to do it and what tools they will use to do it. *How* they are told what to do and *how* they are told to do it matters because workers will intuitively (and quickly) know whether or not the field leaders are serious and truly committed to their safety - just as they will know if their fellow workers are serious about safety.

When we respectfully speak about safety with another person it can produce a real connection because they realize that their safety is of paramount importance to you, the company, and is a first priority on the project. This is how morale grows and so does best safety practice. To be genuine in that regard controls the moral high ground. Anyone who thinks first impressions aren't important is deluding themselves.

Workers and frontline supervisors must make safety personal. We need to show we care about each other. Part of caring power is having the courage to correct a peer.

Attitude of Following the Rules with Flexibility

We know that companies with a strong focus on safety and with constant reinforcement of safety rules have a safer worksite than companies with relaxed expectations. It is very important that everyone onsite has an unwavering commitment to following safety rules and best practices.

Sometimes, however, when workers follow the rules catastrophic incidents can result even when management is sincere. When this happens, workers and supervisors must realize there is something wrong with the system. For example, maybe information is missing or processes are not adaptable enough. What do you do in this case? It is important to ask questions about those rules and the system and understand their weaknesses. Then with new knowledge and answers, update the rules and the system. Then communicate the issue throughout the rest of the company.

In the next example, everyone followed the rules, but there was a problem with the process. Thank goodness it wasn't catastrophic. In this case, the accident was leveraged to make the company's safety policies better.

EXAMPLE ONE: The project was an underwater net repair project located on a remote Alaskan island. An important tool was dropped in ninety feet of water. If the weather permitted, it would take a minimum of two weeks to get a new tool to the island. Luckily, there were three certified divers on the crew. Two volunteered to go down and retrieve the tool. The company provided the diving gear, which included a dry suit (the water was too cold for wet suits).

Buoyancy is regulated in a dry suit by adding air (pushing a button over the chest bone) or releasing air (pushing a button over the right shoulder). One safety regulation that was time honored and inviolate was the use of a flotation vest when you are near or in the water. This applied to divers as a part of a strong safety culture even if they were in a dry suit. Well, you can imagine that it would be very difficult to dive while wearing a flotation vest. The safety professionals authorized use of a safety vest that could be activated by pulling a tab. So if a diver was at or near the surface with heavy gear and needed the floatation, they could activate the vest.

The redundant floatation vest was horseshoe-shaped with a compressed air charge. In an emergency, all a diver would need to do is pull a string and the charge would deploy filling the flat vest with air, which would provide flotation. This type of redundant vest was used during diving jobs all across the company and had a proven track record. All diving equipment on the island was regularly inspected and maintained.

On the day of the tool retrieval, the first diver jumped into the water from the dock by leaning slightly forward, legs apart in a scissor kick - proper technique. But something happened that had never happened before in company history. The impact of her chest hitting the water set off the charge in her redundant vest. It immediately filled with air, which shouldn't have been a problem. In fact, it had never been a problem before because it had always been used with wet suits. But with a dry suit, the deployed vest became a huge problem. As it filled with air, the redundant vest pressed against the worker's chest bone fully compressing the "increase buoyancy button" on her dry suit. In a matter of seconds, her dry suit filled with 300 pounds of pressure - all the air from her tank.

Her feet shot to the surface, looking like small kegs. Her legs looked like blue floating tree trunks so inflated was the suit. The diver bobbed spread eagled on the surface - a cross between a cork and the Michelin Tire Man. The pressure was so overpowering, she couldn't bend her arm to push the release pressure button.

It was quite a sight and she wasn't hurt so they all laughed as the second diver eased into the water to rescue her. The laughter stopped, however, when it dawned on them that if the faulty charge-that-inflated-the-vest-that-filled-the-suit had deployed in ninety feet of water, she would have shot to the surface in a second and died from "boiling blood". For the next three months, the divers had to work against company regulations because it took the company that long to change their rules and regulations and to get the word out to other divers in the company - don't use the horseshoe vest with a dry suit.

In the case of the Alaskan divers, they worked with the project safety lead and decided to order more suitable redundant vests and work without any redundant vest until the new vests came in (which took three months). To compensate for the lost redundancy, they added a person to each dive: two under the water and one on the surface in a boat with a radio - closer help if something went wrong. They also added training so that each diver buddy understood the role of "being the redundant air vest for their partner". They violated company safety policy in order to be safer. In this case, being flexible with company safety rules saved lives.

NOTE: Dayna, one of the authors of this book, was the diver in this story

EXAMPLE TWO: Here's an interesting follow-up. A few months later, on the same project, Dayna was once again diving for the same company. Seventy-five feet under Alaska's frigid water, she realized her thinking had become fuzzy. From one second to the next, Dayna couldn't think. It didn't seem like such a bad thing and she really wasn't concerned. But a refresher dive course lecture worked itself into her numbed mind. "Whenever something doesn't seem right," her trainer said, "get to the surface, and then figure it out. You can always go back under another time. You could have bad air, you could have narcosis of the deep, you could have been stung by something; there are lots of reasons things go wrong. You don't figure it out seventy-five feet under."

So even though they were almost finished with the job (and the job was important) and even though her partner could be mad at her, and even though it seemed overly cautious and even ridiculous, Dayna signaled her partner to surface. Her partner asked why once. Dayna shrugged, but signaled up again. The other diver was well-trained. She didn't like it, but up they went with no questions. In fact, they watched each other closely the whole way up including required depth stops. When they surfaced, the partner asked Dayna what was wrong at the same time she signaled the new redundant buddy in the boat. Dayna spun in a circle, confused and looking for a clue as she slurred, "I don't know what's wrong." That's when her partner noticed the zipper on the back of Dayna's dry suit had broken and opened her to the frigid Alaskan waters.

The boat was right on top of them and the diving buddy and the "redundant boat partner" hauled Dayna into the boat because she couldn't even move her fingers she was so frozen. From the boat he radioed the cook who radioed the doctor (ship to shore) who said to heat up the sauna. Someone on the shore arrived to drag Dayna to the sauna and it was four hours in the heat before she began shaking from cold. It took a few more hours before she could leave the sauna. She lived to co-write this book and she suffered no permanent injury. If she had been wearing that darn horseshoe-vest, the broken zipper would have been hidden from her partner's sight. If they hadn't put the extra person on the dive team as a replacement for the vest, she wouldn't have gotten the medical help she needed in time. If the boat partner had not been fully engaged in his job, Dayna would not have been hauled out of the water, gotten medical advice, or put into a hot sauna in a timely fashion.

These stories demonstrate how individual craft workers have tremendous ability to influence the safety equation by their day-to-day actions, their integrity, and developing their skills, knowledge, and concern for their fellow workers. Other good examples of this include the crane operator who refused to operate the Big Blue crane during windy conditions as well as the foreman who was asked to help a crew member under tight deadline and he took the time to walk fifteen feet to collect and don his safety glasses and ear protection before working - what a great example both men set with their simple actions.

In the case of the Alaskan divers, these stories also demonstrate how important proper training is, especially when the greater mission of safety is kept in focus. These two stories also highlight the fact that when craft workers have the highest view of the company's safety policy, accept personal responsibility and are shining examples of true safety leaders, they can help generate powerful solutions when the rules fail.

Their initiative to red flag the company's procedures was helpful. Their ability to craft a temporary solution that was well considered and approved by local management saved Dayna's life. Their observations about what was actually compromising their safety led to better long-terms solutions. Now understand, we are not advocating going around safety processes and procedures, but to keep the highest understanding of safety in mind! In this case, the temporary solutions worked so well, they became a permanent part of the safety plan. The company's adaptability and training was improved and the company policy was changed to keep the additional person in the boat even when the correct redundant vests arrived.

The greater mission of safety must be kept in mind. Companies often do not have a flexible approach to accommodate improving safety practice. We make note of this in our analysis. Policies must be developed to support such flexibility and teams must have the problem-solving skills to deliver an acceptable solution. On the job, Brion has cautioned his teams, "If you get any kind of gut feeling or uneasiness, stop your operation. Step back and analyze what you are doing." These are words that save lives.

Attitude of Communication

You want to learn. You want to grow in safety knowledge. You want the workers around you to grow and learn, too. Communication is an important part of that. Talk about safety suggestions. Then, if you are a supervisor, act upon the suggestions. This demonstrates to everyone that you care and it builds credibility for your safety culture.

To solicit safety suggestions and act upon them is to show that everyone cares. And don't think you are alone. Get others involved. In order to grow safety culture credibility, you build trust by listening and then implementing good suggestions that are in your scope of authority. Safety suggestions must never fall on deaf ears. Communication is always important.

Communication is a serious responsibility that cannot be taken for granted. Every person must communicate safety principles up and down the company hierarchy in clear and concise ways - constantly. Leaders will be using their safety communication with one main intention: keep people on the safe course.

Does everyone on your crew speak the same language? How well do they understand your instructions? Don't guess. Clear safety communication is too important for guesswork.

A foreman or field supervisor that clearly conveys the vision of the team's capability for amazing safety success will be a source of inspiration. Workers will then rise to meet expectations. When open praise and encouragement are utilized as a way of life, the result is increased self-confidence. Safety success grows and bears the best fruit possible. Such is the power of expectation aligned with excellent communication.

Attitude of Follow-Through

Although it is management's responsibility to conduct formal, rigorous and thorough investigations when an incident or accident does happen, it is everyone's responsibility to participate fully. Sufficient questions must be asked and answered in order to get to the real root cause(s) of any incident, accident, or near hit. Investigators and workers alike must be committed to arrive at the root cause and must find the truth (let the chips fall where they may).

Not all companies have integrity in this area. In some companies, if you admit to a fault, you are fired. Please take a moment and realize where this mindset leads. On the future construction site, this kind of irresponsibility will not be tolerated.

In the current environment, if you work for a company like this, workers have two options. Either quit and find work elsewhere or be a leader and stand up for safety changes. This is a scary suggestion. But personal responsibility is forged in times of stress, not when things are going smoothly. Standing up for safety when there is something to lose is exactly when you build your safety character and commitment.

When an incident happens and management is looking to find the root cause, it is your responsibility to make the project, the site, and your team safer. Do the right thing and fully participate and follow-through.

Safety due diligence requires action and must never be a figment of your imagination.

Attitude of Quality

Developing an attitude of quality is going to be increasingly important as the construction industry changes. You will read more about this in coming chapters, but we want to touch on two quality attitudes that affect safety right now.

Do It Right the First Time

Rework increases risk and costs. Doing work wrong the first time interrupts the schedule, adds delays, adds pressure, and contributes to incidents, near misses/hits, and accidents all across the worksite - not just in your area. It is your responsibility as a frontline supervisor to make sure all workers understand what they will be doing and how they will perform it safely. And make sure the enough time has been scheduled to all for performing the work correctly and achieving these goals.

In order to get your work done right the first time, workers must be encouraged to ask questions if they have any doubts or safety concerns. Getting it done correctly the first time is safer and cheaper!

Refocus Before You Work

Before each new operation, the worker needs to refocus on what they are going to do next and discuss what the risks might be. One of the best tools for this is the use of a Job Hazard Analysis (JHA) for each work operation. We discuss this in greater detail later in the book, but it is so important we will introduce it here.

This means setting the time aside (or plan it into the production schedule) to involve each individual worker and the crew. On this point alone there is no room for talking hindsight. Even though it costs money, schedule time, and reduces short-term productivity, everyone must support this type of proactive action because not only does it help you figure out what could go wrong, it also helps you figure out how to make it go right. And that pays back the time you spent and more.[28]

Be fully engaged when creating your Job Hazard Analysis. Be committed to following it. If your company does not have a formal JHA, get your crew to do it or at the least do it informally before every new segment of your job.

Summary

You don't get to the top by chance, by luck or by accident, no pun intended. It takes bold intent. Ask yourself these questions,

- "Is my level of passion and commitment to safety as extraordinary as it ought to be?
- Am I learning something new every day and pushing to use that understanding to achieve safety perfection?
- Am I willing to face the uncertainty and challenge of dealing with a problem that seems very difficult to solve? "

If honesty causes you to say, "No" then the next question is, "What are you going to do about it?"

Bottom line #5: *Character and leadership are the ultimate expressions of attitudes that make you safe.*

Safety Mantra #5: *Safety begins with me and my attitude.*

Take-Away Points: *Chapter 5*

- Develop strong attitudes and you will grow your character and leadership qualities.
- Safety is everyone's responsibility. Remember "All for one and one for all." Be there for your crew and be relevant to your employer.
- Your safety commitment must be absolute. Be aware that your powerful safety actions strongly influence others.

[28] We will discuss how you can have both improved safety and significant job performance by understanding the processes you are using and where they go wrong. Having a process plan improves safety and the quality of your work.

- Always stop to correct unsafe acts or conditions. Get support if you need it. Be a leader in safety.
- If in doubt, stop working and ask questions. Being 'The Student' is a great attitude regardless of your age or level of experience.
- Always keep the highest safety purpose in mind. Follow the rules, but be flexible and use your head when conditions change. Always follow-through until the issue is embedded in the company best safety practice.
- Over-communicate to engage people and to improve safety.
- When your supervisor is conducting a root cause analysis of an incident, participate fully and honestly. Be sincere in your efforts. Be willing to keep answering, "Why?" Dig deep.
- Do your work right the first time. Before each new operation, refocus on what is next and discuss the new risks. Quality improves safety.
- The journey of safety mastery is like that of an elite athlete. It requires all of your devotion and strength to accomplish.

"Coming together is a beginning; keeping together is progress; working together is success."

- Henry Ford

"It's when we start working together that the real healing takes place... it's when we start spilling our sweat, and not our blood."

- David Hume

Chapter 6 – What We Can Learn From the History of OSHA

Everyone in construction knows about OSHA. We asked many people in the industry to tell us their perceptions of OSHA and they usually say one of three things:

1. OSHA saves lives.
2. OSHA is killing my business.
3. OSHA inspectors are jerks (or nitpickers, or idiots, or... you get the idea).

PERCEPTION ONE - OSHA save lives: Over the decades it has been repeatedly proven that OHSA has saved lives since its inception and will continue to save lives in the future when those standards are adhered to. So this perception is true.

In 1971, the Occupational Safety and Health Administration (OSHA) was established in the United States. Before OSHA, the estimate for the number of workers killed on the job in 1970 was 14,000. In 2009, the number of fatality accidents fell to approximately 4,340. In that span of years, the number of U.S. employed has nearly doubled to 130 million people working at 7.2 million worksites. Further, there was a dramatic reduction in the rate of serious workplace injuries and illnesses. The decline went from 11 per 100 workers in 1972 to 3.6 per 100 workers in 2009. These are striking statistics that demonstrate real gains for safety in the workplace but they are not nearly good enough.

Many states have developed their own OSHA programs and often these requirements are stricter than Federal OSHA's regulations. As well, safety and health professionals, employers and unions slowly but surely have had a significant effect on safety in the workplace. Obviously, there has been an extraordinary commitment to

- eliminating accidents,
- understanding why accidents occur, and
- providing regulatory oversight…

... and yet many accidents still occur. The statistics are sobering. The Census of Fatal Occupational Injuries (CFOI) state that 4,693 fatal work injuries were reported in 2011. This total number of fatal incidences can be viewed from several directions. For example:

- Nearly 1 in 4 of the fatal contractor injuries involved a government project.
- The highest number of fatal contractor work incidents involved falls.
- Pedestrian-vehicular incidents caused the second highest number of fatalities.
- And Hispanic or Latino contractors accounted for 28% of the fatal work injuries, which is well above the 16% share of the previous overall fatal work injury total for that same group.

In other words, what we know and use about accident prevention has achieved a limited success and some sharp reversals. We can take a moment to be grateful for the improvements and ask deeper questions about the reversals.

Could understanding the history of OSHA help our thinking toward future action? The answer is "Yes" because the history of OSHA[29] unveils a history of dedication to worker safety and a history of compromising worker safety at the same time.[30]

We will show how pervasive these basic construction-based contradictions are. Even OSHA falls prey to them.

---

[29] US Department of Labor's website at the history section, which can be found at the following urls: (First accessed 04/2012 and finally in 09/2013) http://www.dol.gov/oasam/programs/history/osha13guenther.htm, http://...osha13stender.htm, ...osha13corn.htm, ...osha13bingham.htm, ...osha13auchter.htm; The Injuries, Illnesses, and Fatalities (IIF) program released two data products on April 25, 2013. The Census of Fatal Occupational Injuries (CFOI) released its final fatal occupational injury data for 2011, which include detailed data on fatal injuries involving contractors being reported for the first time. A summary of final fatal data from 2011 can be accessed at the following url: http://www.bls.gov/iif/oshwc/cfoi/contractor2011.pdf. NOTE: A contractor is a worker employed by one firm but working for another firm that exercises overall responsibility for the operations where the decedent was killed (a business to business relationship not a residential job).

[30] Our mention of "compromising worker safety" is related to the forces involved in forcing the agency to improve its effectiveness from the time it was founded. It is not an indictment of the character of OSHA leaders but, in the author's opinions, a statement of reality that business did not want safety and OSHA made compromises so the agency would survive. Some might reframe this to say that the compromises were actually setting priorities on what mattered most and learning how to become more effective. It could also be a measure of the problem-solving abilities in the business and OSHA relationship. Lessons can be drawn from this dynamic; for example, how easy it is to move to compromise quickly at the expense of informed (problem-solving) examination.

PERCEPTION TWO - OSHA is killing my business: Many people in the industry think OSHA's rules are onerous, overly complex, unjustified, and costly. They blame OSHA regulations for economic hardships. Recent studies, however, suggest otherwise. In fact, it has been shown that random OSHA inspections decrease injury rates, which in turn, reduce medical payouts and lost time/wages costs. In actuality, OSHA regulations and inspections save businesses money.[31]

So, this perception is not true. OSHA as a general rule does not hurt business, so why does this misperception persist? It is important to look at this issue if we want to make safety widespread and systematic on a construction site. Looking at the history of OSHA from a couple of different perspectives may offer us insight into this misperception.

PERCEPTION THREE - OSHA inspectors are jerks (... you get the idea): Anecdotal skepticisms thrive throughout the industry. Just as there are "jerks, nitpickers, and idiots" in any organization, there may be a few at OSHA, too. But we doubt there are any more of these difficult employees at OSHA than there are at your company or anywhere else. The three main reasons you may be skeptical of OSHA inspectors are because you fear them, they may cause you pain (rework, fines, etc.) and because their decisions often seem arbitrary. And that is true.

We need to dig out the fact from the false beliefs. Let's start with a bit of history viewed through an innovator's point-of-view.

History

We can learn much from the history of the relationship between OSHA, Congress, and business.[32] Politicians created OSHA to secure worker safety. But right away, OSHA and business were aggressively at war each other.

---

[31] Levine, David I., Michael W. Toffel, and Matthew S. Johnson. "*Randomized Government Safety Inspections Reduce Worker Injuries with No Detectable Job Loss.*" Science (Weekly) 336, no. 6083 (May 18, 2012): 907–911. (DOI: 10.1126/science.1215191) (Last accessed on 07September2013) - Abstract: Some observers claim that OSHA regulations damage firms' competitiveness and destroy jobs. Others advocate they make workplaces safer at little cost to employers and employees. We analyzed a natural field experiment to determine how workplace safety inspections affected injury rates and other outcomes. Randomly inspected employers experienced a 9.4% decline in injury rates and a 26% reduction in injury cost. We find no evidence that these improvements came at the expense of employment, sales, credit ratings, or firm survival.

[32] US Department of Labor website: "*The Occupational Safety And Health Administration - A History of its First Thirteen Years, 1971-1984*" by Mr. Judson MacLaury, U.S. Dept. of Labor Historian. He wrote this history in 1984 at the request of Thorne Auchter, Assistant Secretary of Labor for OSH. (Accessed between 2012-2013) http://www.dol.gov/oasam/programs/history/mono-osha13introtoc.htm

Businesses held the false belief that taking certain risks was unavoidable (and therefore, acceptable) in order to earn a profit. They claimed that in order to get a job done, craft workers must occasionally face death or serious injury - it was the nature of the work. Business's bias was to improve profit over worker safety and they lobbied Congress constantly. On the other hand, OSHA held the false belief that since industry created their own voluntary guidelines that holding businesses to those standards would be welcome oversight as a basic minimum.

This led to dramatic political clashes and harsh criticism of OSHA's "interferences". Congress declared that OSHA had insignificant accomplishments in their first ten years of existence as an agency and said that OSHA had earned a reputation of being a "nit-picker" because of their strict adherence to "some silly things" and because of their preoccupation "with frivolous, irrelevant rules and regulations." The political scene, during these first OSHA years was turbulent.

Dramatic accidents put heat on Congress to revamp the agency. It was obvious that voluntary compliance was not working. Congress demanded the agency enact reforms but not be too dictatorial. OSHA's leadership created a more efficient operation and decisions were based on science. They also simplified the standards and requirements so businesses could comply with OSHA's guidance more easily. These milestones plus two others helped change the relationship between OSHA and industry:

- The ability to create an "equivalent solution",
- OSHA focused on hazards and went back to an enforcement model for a pattern of indifference to worker safety.

Before "equivalent solutions" existed, companies were required to exactly follow standard protocols in order to be deemed safe. But sometimes a company came up with a safer way of performing their work. If this safer process didn't follow the standard, however, it was not considered safe by OSHA's standards - even when it was safer. The company could be punished for non-compliance even though they were achieving better results than the standard demanded. No wonder there was industry and political backlash. Equivalent solutions meant OSHA could set standard minimum safety regulations, but it allowed them to accept any solution that had equivalent safety results. This is when the dramatic safety and health results produced by OSHA began.

OSHA still has a focus of enforcement on companies with a history of violations in both federal and state plans and has been very assertive regarding a pattern of indifference to regulations and worker safety. Since 2008, if a specific company site is in violation, OSHA may seek company-wide compliance as part of the resolution process. Penalties can be very stiff for those who are indifferent to their obligations. And the agency works for increased whistleblower protections for employees. OSHA knows that business can do better.

These have produced safety gains, but there is always a sense that OSHA is interfering, too. And that conflicted perception shows there is still a problem in the industry.

The Relationship Value Ladder [33]

A robust safety environment requires cooperation, mutual respect, process understanding, and problem-solving skills. One of the reasons OSHA has seen so much trouble since its inception is because its relationship with the business world has been based on opposition and not upon mutual respect and cooperation.

Briefly consider your personal relationship to OSHA, safety standards, and your safety relationships in construction. If you are thinking

- OSHA is costing our company,
- Safety inspectors are idiots,
- Safety initiatives are getting in my way; please, just let me get my work finished...

... then you can improve your company's bottom line and safety performance by making better relationship and attitude choices.

Obstructive conflicts hurt safety performance and profits. Mutual respect and cooperation improves safety performance and profits. This is a great and underutilized opportunity that permeates construction. This is not a sweet little platitude; it is an important differentiator if you leverage it correctly.

Some problems are easy to solve. They can be solved the same way they have been solved in the past or they can be obvious, next-step answers. Not every problem needs an inventive breakthrough. Not everything needs to change. Most aspects of your construction culture are precious and should remain just as they are.

Having said that, everything needs to be examined before you decide what will remain the same and before you determine how you will solve your issues. Everything needs to be assessed!

33 The Relationship Value Ladder, the Four Relationship Levels and their outcomes are based on work that Innovation Scientific created through the analogical comparison of Safety and Quality and the associated behaviors and responses that occur at each level. The concepts for the Four Progressive Levels of Quality were also created by Innovation Scientific to show how a change in mindset is essential to move from inspection-based work to process-oriented work.

This includes construction relationships. The construction industry is very traditional, which means the relationship momentum is strong and largely unexamined. You relate the same way you have always related to the people and organizations that you have always interacted with. In some ways, many of your construction relationships are "in cruise control" and up to 85%[34] of your relationship responses are probably automatic and based on how your were trained.

This is fine... unless there are problems within your organization that involve compromises and trade-offs. When these troublesome conditions exist and you don't know how to resolve the issues then it is time to turn to new resources. New resources include new relationships (or old relationships used in new ways).

When the goal is complex or tough, it may be time to climb the ladder of potential relationships. Take a look at your key industry stakeholders.

You probably have ideas about them. Pay attention to what passes through your mind and emotions as we list each one.

- OSHA
- Co-workers
- Management
- Unions
- Insurance companies
- Suppliers and vendors
- Lawyers
- Your competition
- Subcontractors
- Clients (at certain times)

If you are honest, many of these relationships are probably antagonistic. Relationships often begin with antagonism. Think of this like a ladder. The easiest rung to access is the first step. It is also the least useful in most cases. Antagonistic relationships are the bottom rung of the relationship ladder.

When you climb the relationship ladder, you move your aggressive relationships into easier but still reluctant voluntary interactions. If you keep working at them, they escalate again to mutual cooperation. And finally, if you keep climbing the relationship hierarchy, you reach mutual voluntary engagement (win-win).

Each of these steps up the relationship ladder gives you more resources to use to solve your entrenched challenges. Once you start moving towards zero harm on the

[34] This number is approximate and based on a stimulus-response model of human behavior.

job (without compromise) you will realize that everyone must benefit from safety solutions and, thus, your relationships will need to evolve to support this.

Right now it seems impossible that OSHA and businesses would help each other be more productive and safe; seems impossible that you could trust each other or have each other's back. In fact, when you imagine it, it probably feels wrong - even unnatural. Maybe it even makes you feel sick or angry.

But the truth is, in the future you will have top-level relationships with some of the organizations you would never have dreamed of before.

- Or at least *parts* of your organization will have win-win relationships with *parts* of OSHA's.
- Or *some of the time* your company will act synergistically with theirs.
- Or maybe *in specific situations and under certain conditions* you will be allies with organizations you cannot imagine now.

These partnerships are currently unlikely but they will develop eventually because the future we are envisioning will demand it. In order to make the safety gains we are writing about, you will need to have relationships high on the relationship ladder. If the industry is going to be more systematic, safety-pervasive, scientific, and process-oriented, then it would be good to have government organizations footing the bill for some of the industry-wide research and development. Also, if OSHA is involved, they will probably have free or low-cost training for craft workers and frontline supervisors so you can position yourself at the head of the safety employment queue.

After reading this book, we hope you will have developed a flexible mindset so that you will be able to envision and competently use not-obvious alliances when everyone else is stuck in their traditional relationship point-of-view. This can give you a great competitive advantage.

Inspection-Based Safety and Production Models Are Countermeasures

Businesses still believe that compromise about worker safety is essential because no more can be done to advance safety while OSHA firmly believes that more can and should be done and is recently getting stricter on enforcement. One of the ways to advance up the relationship ladder is for the construction industry to advance from the crafts model of production to a modern-model based on science to improve its safety and production performance simultaneously. The crafts model of production is limited in its current state:

- It is based on the heavy use of countermeasures and not real solutions;
- Inspection-based compliance is essentially antagonistic.

Think carefully and realize that safety inspections are countermeasures. They are not solutions. So right off the bat, they use a lot of resources (time, money, etc.).

Inspection results are based on judgment calls and every inspector may notice and call out different things. In other words, they are not standardized; they are not interchangeable; they are not consistent. Instead, they are arbitrary. That is one of the reasons people think OSHA inspectors are jerks; they think they are being arbitrary. Businesses feel they can never prepare adequately for an inspection since they don't know what that particular inspector will notice.

For a moment, let's reverse this thought. Your processes are arbitrary. Your safety practices are random. Your worksite is unstructured and that makes it very difficult to inspect. Many non-construction industries can prove that their processes are stable and in control and their safety problems with OSHA are small.

Inspections Are the Lowest Form of Assurance

Inspections are performed in real time as the work is being produced, which means they can detect mistakes only after they have been generated (or when they are in the process of being made). Inspections are not predictive. They are not systematically integrated into decision-making and they are not scientific. They catch infractions; but don't stop hazards from being created in the first place. They are better than nothing, but mistakes will be made because there are not enough inspectors to be everywhere at all times and no human can be 100% vigilant.

Inspections are at the lowest rung on the relationship ladder pitting one side against the other.

Inspections will not assure that your workers are safe. In fact, inspections are the lowest form of assurance that can be provided. Remember, a countermeasure requires constant reinforcement, redundancy, maintenance and vigilance to be effective and a countermeasure never actually fully solves the problem. Inspections, therefore, must be constant, omnipresent, vigilant, and redundant to be effective. But they are not constant; they cannot be omnipresent; and it is too costly to be 100% vigilant and redundant.

Inspections cannot detect unexpected human blunders. Despite rules and standards and regulations workers will sometimes fail to perform a required action. Sometimes workers will mistakenly perform a prohibited task. And sometimes there are communication gaffs and misinterpretations of essential information. Unless the inspector is right there as the blunder is happening, there is no chance of stopping it by inspection.

Because inspections are countermeasures they are an ongoing cost. The cost will never end. However, if you generate solutions this "never ending cost" savings is the money that you get back from your investment. If you develop these solutions it helps protect your job.

Four Levels [35]

There are multiple metrics we have defined that can help construction companies understand how they are choosing to relate to OSHA (and unions, and co-workers, etc.) and the resultant expectations and outcomes of each level. Although we are applying this to safety, it is much like what happened in the quality movement.

	Science Level	Prevention Level	Relationship Level	Behavioral Level
4	Multi-disciplinary Science / Inventive Thinking / Predictive Statistics	Standardize and control key process drivers of inter-related processes. Inventive solutions.	Mutual engagement	Respect and contradiction resolution. Win-win strategy.
3	Science / Predictive Statistics	Standardize and control key process inputs. Solutions.	Mutual cooperation	Contradictions in relationship. Some win-win.
2	Logic / Statistics	Standardize and control process outputs. Countermeasures.	Reluctant cooperation	Viewed with suspicion. Win-lose / lose-win.
1	Judgment Call	Inspection. Countermeasures.	Antagonism	Open Hostility. Lose-lose.

The Four Levels and Their Outcomes

Assurance Level One: Inspection; **Level One Relationship**: Antagonism between inspection and production goals.

Assurance Level Two: Standardize and control the process outputs; **Level Two Relationship:** Some cooperation along with the suspicion between inspection and production.

---

[35] The Relationship Value Ladder, the Four Relationship Levels and their outcomes are based on work that Innovation Scientific created through the analogical comparison of Safety and Quality and the associated behaviors and responses that occur at each level.

Assurance Level Three: Standardize and control key process input drivers with predictability of process outcomes; **Level Three Relationship:** Mutual cooperation but with many contradictions in relationship.

Assurance Level Four: Standardize and control key process drivers of all inter-related processes system-wide including management and project delivery with predictability of process outcomes; **Level Four Relationship:** Mutual engagement, respect, contradiction resolution, and win-win strategies using a multi-disciplinary and scientific approach.

Personal History with OSHA

Some of you have direct experience with OSHA and may even have contradictory feelings about the agency. You like that OSHA keeps you safe, but you would hate it if an OSHA inspection led to you getting fired.

EXAMPLE: a foreman had his crew install rebar caps. The crew was behind schedule and there was a lot of work left to do. Since he was in a hurry, the foreman didn't check to see that all the rebar were capped before hustling his crew to a different part of the project. An OSHA inspector held a surprise visit and found a few uncapped rebar in a dangerous location (if someone tripped and fell, they could have been impaled). The inspector levied a fine against the contractor. The contractor, in turn, fired the foreman. The foreman resents the OSHA inspector to this day but begrudgingly is grateful the inspector caught the mistake before anyone was hurt. "The way we were running around that site," he says, "someone would probably have gotten hurt sooner or later."

This foreman is more careful now to double check that all rebar is capped on his projects and he does appreciate OSHA's protections. However, he resents that his job may be in danger if he makes a mistake. He realizes that OSHA inspectors are there for worker on-the-job safety but they are not there for the worker's economic well-being.

OSHA is a treasure-trove of safety knowledge and an amazing resource and yet you shake and shiver in their presence and finger-point when they leave. Based on what you have already learned in this book, what does this tell you about your mindset?

What can you do? OSHA can only set the minimum standards for safety. If you are to achieve more you must take the lead away from OSHA and heavily leverage the equivalent-solutions rule. You must change how you utilize OSHA so that you can get what you want - a safe workplace, get your work finished on time and keep your job.

We keep insisting that you will need to be more proactive and responsible in the future. Here is a hint for those of you who are looking for a future advantage. Check out OSHA's *Field Operations Manual for Inspectors*. Even though it's written for inspectors, it's available to anyone for free on the OSHA website (you can download it). It's a great resource to help you gain a competitive personal advantage in the marketplace while you help your company prepare for the possibility of an OSHA inspection.

Lessons

So why spend the time analyzing the history of OSHA? Well, some lessons emerge that help us understand what workers and frontline supervisors need to know.

Anytime the project pits safety against profits, schedule, or productivity, workers and supervisors must address ALL sides of this challenge. Choosing worker safety *without consideration* for business profit (or visa versa) means neither side gets what they want. But on this important subject, compromise is not the answer either.

- LESSON: In other words, great solutions have a bias toward safety but address the needs of both sides simultaneously.

Just as OSHA needed to become less frivolous (irrelevant rules and regulations) over time and learned to make better use of resources, so do you.

- LESSON: This applies to workers because you cannot expect to do the same job and deliver the same kind of results and remain relevant. You are a resource for the companies you work for. You must keep adding more value over time. No one is indispensable unless they continuously make themselves so.

OSHA's rules and regulations were complex because they needed to create countermeasures for the lowest common denominator. As they became better at their job, they simplified regulations and standardized rules.

- LESSON: This will be true for workers, too. As the new systems and sciences are put into place there may be more rules you need to understand and implement. However, as processes and policies become standardized, things will be simplified. If you want to keep your job during this transition, you will need to be flexible, adaptable and will need to continue learning.

Sub par safety standards do not lead to 100% safety success. Neither businesses nor workers can just comply with OSHA's minimum safety regulations and expect great safety results.

- LESSON: You cannot accept different safety standards on different sites; the safety standards you hold for yourself must be consistent and pervasive.

The relationship between key stakeholders is evolving from antagonism to voluntary limited participation. Currently the construction industry is between rung one and two on most of their relationship ladders.

- LESSON: If a situation is antagonistic, move the situation towards limited voluntary participation, then to full cooperation, and finally to *mutual engagement.* Be the team leader in this area.

And finally enforcement must be strong against a pattern of willful disregard for worker safety.

- LESSON: Never pass an unsafe situation or action without correcting it. If you do, expect to be called to task.

Bottom line #6: ***New levels of safety and profit performance will be achieved through mutually engaged stakeholders striving for win-win solutions, resolving contradictions, and using a multi-disciplinary and scientific approach to safety.***

Safety Mantra #6: ***Win-win is created when we solve tough safety problems together.***

Take-Away Points: ***Chapter 6***

- Drive fear out of the organization.
- OSHA regulations and inspections save lives and money. But a process focus and scientific approach is a better way to save lives and make even more money.
- Safety inspections are countermeasures, not solutions, based on judgment calls. They can only detect mistakes after they have been made.
- Advance your safety knowledge to be more scientific, process-oriented and preventative to make inspections unnecessary.
- Great solutions have a bias toward safety but address all priorities simultaneously.
- Craft safety solutions that are relevant to the priorities of all stakeholders. This is an essential prerequisite to having solutions be fully accepted.

- Engagement and mutual cooperation create better solutions. Obstructive conflicts hurt safety performance and profits.
- Operating with an attitude of antagonism among key stakeholders creates solutions that do not last.
- You remain a relevant resource for your company when you deliver better results.
- You cannot accept different safety standards on different sites; the safety standards you hold for yourself must be consistent and pervasive.
- To move your safety agenda forward, you will need to be flexible and adaptable and grow your process, problem-solving, and safety knowledge.
- Learn from the past so as not to repeat the same mistakes. This is essential to creating safety solutions. Use an inventive analysis of historical decisions to anticipate and to avoid future safety issues.

"The common thread between all great innovations is the resolution of contradictions."

- Genrich Altshuller, Founder of TRIZ

"Smack in the centre of contradiction is the place to be."

- Bono

Chapter 7 – Safety Contradictions / No Compromise

This is one of the most important chapters in this book. It is also a bit technical and we are going to walk through it carefully. If you understand this chapter, you will grasp something very, very necessary in the coming future. Prioritize the time to own this information. If you have questions, share this book with others and have a group discussion on this chapter. Get training if you need to. This chapter could change your life.

Very Quick Review

Companies, managers, and frontline supervisors want to put people's safety first and integrity demands that safety is never compromised, but reputations are built or lost on the ability to bring a project in profitably and we also want to protect jobs. Sometimes it seems that we can only get what we want at the expense of another important value.

By now, you may finally admit that there is a huge contradiction between profits/schedule and safety in the world of construction. And if we are honest, we have all bewailed the fact that we can't have both optimized profits and 100% safe projects.[36]

Accepting this kind of trade-off means you have a *mindset of compromise*, which leads to accidents. This is why virtually all companies cannot eliminate that last 5% of accidents no matter how hard they try. It is easier to compromise rather than solve

[36] Any habit of resentment and any automatic response of resistance to safety means you still have a hidden priority towards trumping safety in the pursuit of greater profits. By now, every time you feel these emotions, you need to bring them from the shadows out into the light and notice that this is your reaction. It's a solid step towards being safer and more profitable.

these contradictory problems. People choose to trade-off or compromise more often when they are under pressure. These trade-offs are often countermeasures.

What we need to do is resolve the contradictions and have the best of both sides of the compromise. Certainly a person with an inventive mindset would work towards having the best of both. And the construction worker of the future is going to need to be adept at resolving contradictions without trade-offs.

Mixed Messages Reveal Contradictions

Contradictions often appear as mixed messages.

EXAMPLE: A welder was called off of his current work to drive to the other side of the construction site to solve a little welding dilemma. His boss told him to go, but hurry back as the welder was needed for his current task. *[This was the first mixed message.]* When the welder showed up, the supervisor, foreman, and the rest of the crew clapped.

"Finally you're here," the foreman called out. "We're way behind schedule," he continued while the supervisor nodded in agreement. "Someone was supposed to put this safety rail in yesterday. We're all waiting to get to work."

Some on the carpenter crew made good natured jokes about how welders had the easy schedules. "Welders get to sleep in and take long coffee breaks and go home early," they laughed and elbowed each other. "We were slaving away at the crack of dawn yesterday and had to stay on the job until the day's work was finished and we have to stay late tonight to make up for this gaff." *[This was the second mixed message. Some welder had blown it, which is a "you' versus 'us' attitude; welders don't work as hard as carpenters; and since it was that other welder's fault, this welder was responsible to make amends.]*

The welder was told he needed to construct a temporary stair rail of angle iron and wire cable so the woodworking crew could get up to the second floor safely. He looked at the situation and realized he did not have everything he needed in the truck. The supervisor kept glancing at his watch. A few men made more jokes. "Wait," he said, "I think I have a work-around." He set up his welding equipment and waved everyone back. *[This was the third mixed message.]*

Soon sparks were flying as the welder attached a long piece of welding rod as a structural element to the stairway, which is an extremely poor welding practice. The bead was intermittent, too. He knew the job was not safe and he really needed to start over with the correct materials, but he was embarrassed to say so and did not do it. *[This was the fourth mixed message.]* The supervisor saw the welder's embarrassment and thought the welder was ashamed of the weld appearance. The supervisor did not know it was unsafe and told the welder not to worry and get back to work. The rest of the crew went up the stairway structure carrying loads of materials and eventually the attachment weld failed and someone was hurt.

Mixed messages cause confusion and carelessness. These mixed messages are due, in large part, to an ignorance of the importance of contradictions. For example, if the welder had been trained in recognizing and solving contradictions, he would have realized that:

1. The welder's boss really did not want him to go, but told the welder to leave for the other job *[the first contradiction]*.

2. The schedule was in conflict with his safety best practices. Someone else's oversight caused peer pressure for him to work too quickly *[the second contradiction]*.

3. His own unease was pressuring him to make an unsafe choice; he knew it was unsafe but made the choice anyway *[the third contradiction]*.

4. When the supervisor misunderstood the welder's embarrassment, the welder did not correct him. The welder allowed himself to be sent away. He felt relieved of responsibility - he was just following orders *[the fourth contradiction]*.

During this series of mixed messages the welder felt pressured and in that situation could not think of a solution that would make everyone happy. Because he thought these contradictions were impossible to resolve, he chose to lower his commitment to safety for 'practical reasons' just this once. It resulted in an accident.

Mixed messages cause confusion and carelessness. They should not be delivered and they it's best not to ignore them either. When faced with mixed messages, what do most workers do? Most workers either adopt a similar mindset of compromise[37] or they become angry at the company for not addressing their perception of what is important, but say nothing. Neither of these responses is appropriate.

Because our most reliable safety tool is a passionate person who is good at their job and has safety on their mind continuously, that is what we must demand from ourselves. This (our best safety tool) is a countermeasure, which means each person must constantly reinforce proper safety measures, they must have redundant safety back-up plans, they must constantly maintain vigilance to work safely. To an innovator, however, these mixed messages carry an understanding that something

37 The origins of the word 'Compromise' is Latin and Middle French, which means "together promise". The implication is of joint cooperation and some sacrifice for both sides to reach a solution. Instead of looking at sacrifice we are looking to create a new implication for win-win solutions in safety that fulfills a promise for everyone with no sacrifice.

fundamental is contradictory in the system. Mixed messages must be decoded into a business contradiction and then be resolved. Every worker must try to achieve this level of clarity and insight. For example,

1. When the welder realized that his boss really did not want him to go, the welder should have taken a breath and decoded the mixed message. "If I understand you correctly," the welder could have said to his boss, "both welding jobs are important. You want me to help that other crew but you also want me to get this job finished for our crew. Which one is more time critical? Is there another welder available to handle either of these projects?"

2. The schedule was in conflict with the welder's safety best practices. Someone else's oversight put pressure on the welder to work quickly. The welder needed to take a moment to decode the mixed message. He could have said, "You're holding me accountable for *your* welder's blunder but I'm not responsible for the delay. I need your help. Instead of you guys standing around and pressuring me, can one of you drive the truck and get this list of supplies for me while I set up my equipment? I need to have the correct materials so this stairway is strong enough to keep you safe. Who has the right license to drive the company truck?"

3. His own unease was pressuring him to make an unsafe choice; he knew it but did it anyway. The welder needed to take a moment to decode the mixed message from his own self doubt. Internally the welder needed to reaffirm, "Safety is most important. I want to do a good job and keep these guys safe. But, they are behind schedule. How can I use them as a resource to help get this done more quickly and build teamwork?"

4. When the supervisor misunderstood the welder's embarrassment, the welder did not correct him. He needed to take a moment to decode the mixed message from his own fear. He needed to say internally, "Right now I have the lead to keep this crew safe. And I never walk away from an unsafe action or condition without taking action!"

So here is a quick summary about mixed messages. Don't ever give them. Come right out and acknowledge the unresolved contradiction within the situation. This builds trust and provides opportunity to advance people's safety. Even if the contradiction cannot be solved at a given point in time, if it is brought into the light it can be solved over time. This will reduce the level of compromise inherent in the situation and will continue to improve credibility and build trust.

When leaders compromise safety, everyone who observes them will begin to doubt the leadership. Despair, serious attitude problems, and a loss of confidence can be the result when a leader fails to walk the talk. That is a recipe for eventual disaster, which could be avoided. The long and short of it is that there can be no mirage, smoke and mirrors, or faking it. No matter what leaders say, it is their day-in and day-out priorities that scream the loudest. Fellow workers can see right through any lies or baloney, hypocrisy or ignorance.

The best tools we currently have for construction safety are the *consistent hammering* of safety messages, *unrelenting vigilance*, and *integrity*. Now you have new tools, too:

- Identify and eliminate things that increase risk, like unproductive fear and ignorance.
- Think more clearly and understand the safety issues at a deeper level.
- Think in terms of contradictions so you can identify all the things you want that are hidden in communications.

Use current best practices and if we are lucky, that will keep us safe 95% of the time until new tools, knowledge, and skills can be developed.

And remember that contradictions often appear as mixed messages.

Contradictions Point to Problems and Also to Opportunities

Words such as "Contradiction", "Compromise", "Optimization", and "Trade-off" should set off our problem-detection radar. When we hear any of these words, we need to understand a hidden opportunity is present.

- Contradictions are Opportunities
- Compromises are Opportunities
- Optimizations are Opportunities
- Trade-offs are Opportunities

We hope this is clear because the contradiction/opportunity mindset is going to be a big part of the change coming to the construction industry. It is a huge opportunity for you to add value if you understand this chapter.

So here it is again, said another way. If you have a problem and you can describe that problem in terms of a contradiction, compromise, optimization, or trade-off, then it can be solved. It is that simple. If you can see the issue as a contradiction, it is solvable.

What Is A Contradiction?

A contradiction usually makes you think of compromising. A contradiction is a problem that makes you choose what you want most. For example, I want more money and I want more time with my family. If I work overtime, I'll earn more money but I'll have less time to spend with the kids. If I take time off work to go see the kid's play, I'll have a smaller paycheck. What is more important: the money or the time with family? I have to sacrifice one in order to have more of the other one. That is a contradiction. It's a sacrifice, a trade-off, and a compromise.

But, what you really want is more of both time and money - and that seems impossible.

When a problem seems impossible to solve, that is when most people decide to compromise, trade-off, or sacrifice one priority for another. It takes a unique mindset to face a problem that seems impossible to solve. You have to be tough to tackle a seemingly impossible problem. It's scary. It's overwhelming. It's hard to believe that you don't have to make a choice about priorities. The whole world view is set up to force us to choose one important thing over another important thing. Some of the most intimidating life decisions we face are contradictions.

We can all agree that if we were good at resolving contradictions we would have a significant competitive advantage in life.

On each side of a contradiction is something we want. Typical construction contradictions we might face include:

- Quality vs. Cost
- More Strength vs. Less Weight
- Higher Qualified Workers vs. Lower Payroll
- Staying on Schedule vs. Adverse Weather Conditions
- Profit vs. Quality Materials
- Bidding Low Enough to Win the Job vs. Time Enough to Perform the Work Safely

The list goes on and on. And, of course, let's not forget

- Production versus Safety

You, the worker and frontline supervisor, must understand how to deal with contradictions because they are going to become increasingly important in the future. In construction's future, a large part of project management will be contradiction management. Solving contradictions will be a basic skill that every worker and every level of management must understand. The ability to solve contradictions will become more important as the industry grows and changes. This is one way you can really add value in the changing construction world.

Contradictions are Opportunity

So far, we know that contradictions are major red flags. They signal us that inherent in this conflict is an opportunity waiting to be set free. Michelangelo, the famous architect, painter, sculptor, and poet gives a great analogy when he said, "I saw the angel in the marble and carved until I set him free." Just so, we see beautiful safety opportunities in contradictions and we problem-solve until we set it free. You could do the same and add real value.

Complex and Simple Contradictions[38]

How do we solve contradictions so that we get more of what we want and remove those things we don't want? The first step is to recognize that contradictions come in two flavors: complex and simple.

Most people cannot solve complex contradictions because they are too difficult. Instead, they prioritize one factor over the other. Falling into a trade-off mentality is too easy.

How would you solve a complex contradiction such as cost vs. quality? Where would you start? It's too big. It seems impossible to get cost down and quality up (unless you sacrifice time - and that is a trade-off, too, so it is off-limits as a strategy).

You could use science and math to help you, but using innovation to solve contradictions is much easier and faster. Here's how you start. Complex contradictions need to be broken down into more relatable ideas. In order to solve a contradiction, we must transform the complex one into a simple one.

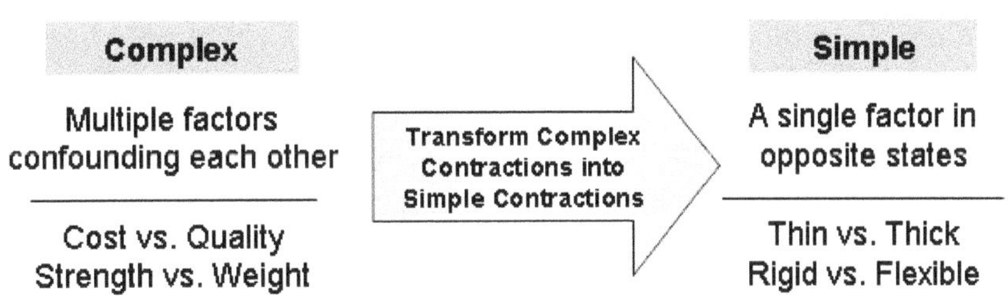

First Step to Solving a Contradiction, Transform Complex to Simple

Break a complex contradiction down into many simple contradictions, which are much easier to solve. For example, a simple contradiction uses a single idea

[38] Dana Clarke, Sr. of Applied Innovation Alliance redefined the TRIZ terms, technical and physical contradictions, (now called complex and simple) for the Structured Innovation Toolset.

(thickness). It must be thick in some places and thin in other places. Or, it must be thick some of the time and thin at other times. Or use another simple idea (rigid). It must be rigid in some places but flexible in other places. Or it must be rigid some of the time but flexible at other times.

Let's look at the solar panel example in the figure below.

EXAMPLE ONE: The solar panel needs to be sized correctly to deliver enough power. But when we increase the panel-size the shipping cost also increases and we don't want that. Due to their awkward size, large solar panels tend to get damaged very easily, too. This is another thing we don't want. At first glance, the 'panel-size versus cost-and-damage' contradiction is difficult to solve without trade-offs.

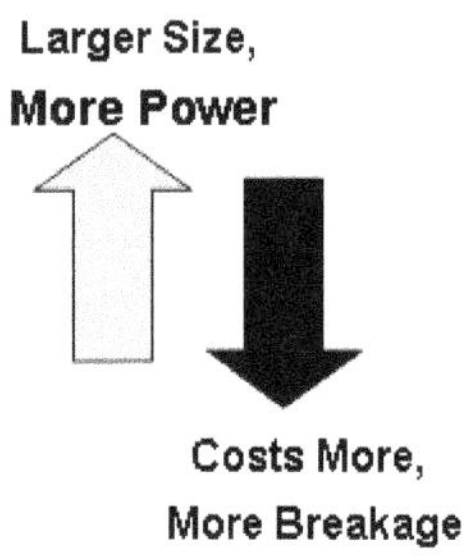

Two (or more) Parameters Opposing Each Other

- Power versus Cost
  - As the panel size increases (improved performance/power output – what we want);
  - However, shipping costs and panel handling breakage costs also rise (not what we want)

A Complex Contradiction

But notice what happens when we turn this into a simple contradiction. To convert it, we start stating reasons: To get more power we need the panel to be larger but that means it will cost more to ship and it is more awkward to handle, which increases breakage. Wow, right there we have a simple contradiction - size. In some cases the size needs to be big and at other times it needs to be small.

We need the solar panel to be both large and small. We need it to be large when making power and small when handling or shipping. *(See the figure below.)*

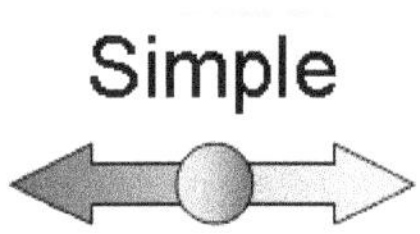

- **A Single Factor In Opposite States**
- 'Size' is a single factor
  - We need the size to be large when we are generating energy
  - We need the size to be small when shipping the panel
  - We need the size to be small when handling it

A Simple Contradiction

State the reasons why we need a large panel size and a small panel size. We need the panel to be large for power output but small for reduced cost (ease of transport, improved safety, less damage, and lower shipping costs). We can see that both values are important.

NOTE: *Notice how some unexpected benefits we get for free (safety-easier to carry and less panel damage) start to show up when we solve for the original problem. This is true of resolving many contradictions.*

As soon as we state the contradiction in this way (with a single *simple* parameter in common with both sides of the complex contradiction) it somehow seems easier to solve. Did you start thinking of hinged and foldable solar panels? When it is open all the way, it is large enough to generate the required energy. When we need to ship or handle it, we fold it smaller. That is just one solution, of course.

EXAMPLE TWO: Now let's discuss a historic example from the world of construction: the I-Beam.

Builders have always needed materials that are strong yet inexpensive and less heavy. That is a complex contradiction. But without the discipline of identifying contradictions, the brain defaults to the common trade-off answer: if you want more strength then you need more material. But more material is costly and heavier, and that's no way to build a skyscraper.

But, if you state, "We need the beam to be strong along its cross-section height but it only needs to be half as strong along its cross-section width. If I change its shape to be tall in its cross section and narrow in its width it will be strong where I need it. That way we could have all the strength we need but less material overall, which makes it cheaper and lighter."

At this point, you can see how it would be pretty easy to come up with the I-Beam solution. Later when engineers discovered much of the strength in a beam occurs by the beam's shape and not by the amount of material, hollow sections came into being.

This thinking process is very effective when facing any contradiction - even safety and risk challenges.

First, you must realize that our ordinary thinking process is often non-specific. If we want something to be stronger, then we say, "Give it more material." It is only later we realize what the secondary problem is: more material also increases costs and we can't seem to have both more strength and low weight.

If we examine our problem we may notice that we only need the strength under certain conditions; maybe certain parts of the system need to be strong; maybe spatially we need strength in a specific place, or maybe it only needs to be strong at certain times.

In the I-Beam example, we are allocating more material in certain places and less in others. This adds strength in a certain place as stated above.

By separating that one single parameter (material) that is in common with both strength and cost and by strictly allocating the resource, we solved the problem with that simple parameter. Solution: One single parameter (more material) in some places for strength and removed (less material) in other places for cost.

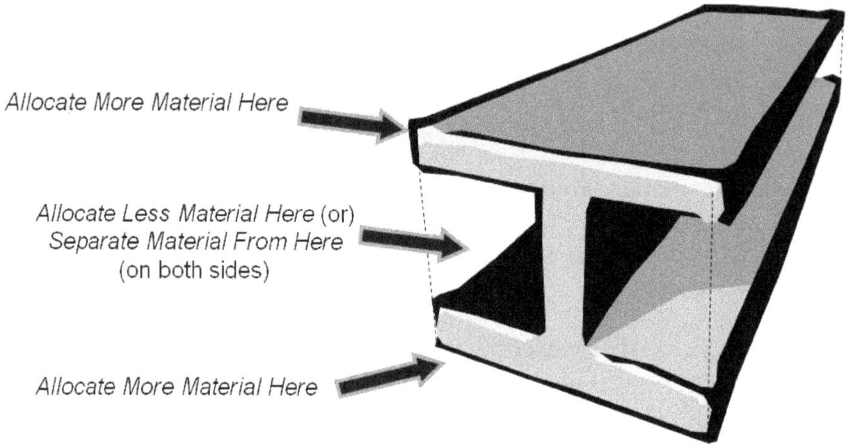

I-Beam vs. Solid Beam: Separation Principles or Allocation Principles

There are other innovation tools that make solving contradictions even easier. You can look into Structured Innovation or TRIZ training if you want to learn more.

EXAMPLE THREE: In construction, cranes often need to work in the same area at the same time. That's a complex contradiction. We can turn it into a simple one if we realize that *height* is a common factor. Two cranes can work in the same space if one is set to work from 14 -22 floors up and the other one is set to work from 0 - 13 floors. Or we can use *time* as a common factor. Software manages cranes in time (only one crane and its load can occupy a specified space at a time) or *upon condition* (load picks have priority; resting cranes need to yield to moving cranes). Of course we are trying to avoid the fatal contradiction of two cranes or their loads occupying the same space at the same time while performing work, which can equal an accident.

This thinking process is very common. Our brains are built for solving contradictions. Almost all adults can do it. But our human brains like to be resource efficient and it is easier for our brains to default to the simplest thinking process (trade-off and compromise) first.

Think of it like aerobic training. Your brain notices that you exercise your heart and lungs by running every day. Then you go on vacation for a week and lay by the pool drinking umbrella drinks. When you get back to your normal routine, you are bummed to notice that you've lost a significant portion of your aerobic training and have to work very hard to get it back. That's because the brain and body like to be resource efficient. It decided if you were not using those muscles in peak performance, it would lose them. That is just like solving complex problems (contradictions).

You need to practice solving contradictions all the time in order to train you brain. Solve easy ones; solve hard ones; make complex contradictions into simple contradictions. Notice when you want to use a trade-off. Notice all the contradictions that have already been solved in your daily life.

For example let's examine an easy contradiction that has already been solved for thousands of years: on a construction site, accidents happen on roads. People need to drive forklifts going to the west side of the site at the same time a worker needs to drive a forklift going east.

"What was the contradiction?" you ask yourself to train your brain. "Multiple vehicles need to use the road at the same time but go in different directions. Before this problem was solved, it seemed impossible. They would crash and they did crash."

The next step is to practice turning this complex contradiction into a simple one by asking yourself, "What is in common with this contradiction?" *Space* is a common factor. Roads are divided in half lengthwise and you can only drive East on this side and can only drive West on the other side. Practice turning this complex contradiction into a different simple one by finding another common factor. How about *time* as a common factor? We put up a stop sign at an intersection, which lets first one vehicle go through, then lets another go through. Here's another common factor. When we notice a person directing traffic and she signals the dump truck to stop and the forklift to go, then we are separating traffic upon *her signal* and *time* together. The dump truck goes when the conditions are favorable (when she signals that it is the truck's turn). See how easy it is to practice.

Whenever you think, "I need to compromise," make note of the compromise and make note of your mindset and the triggers that cause you think of trade-offs. Then when it is safe, find some quiet time and practice solving the contradiction. You need to exercise the "contradiction-solving muscle" in your brain[39].

---

[39] Obviously it's not really a muscle; it is really a series of chemical interactions and electrical signals firing across synapses that are built and strengthened through the use of innovation (problem-solving) processes.

You must practice seeing the world as complex and simple contradictions. Pretty soon you will get very good at seeing the world this way and will know when to compromise and when to solve the problem. At this point, you will be managing contradictions. You can create a laundry list of safety problems that you want to solve. You will become very valuable to your company if you can save them money and/or time while making people safer.

Once you get good at it, you will be shocked at how simple it is to solve problems other workers think are too hard to tackle.

In summary, you resolve contradictions using a structured thinking process that creates solutions using common high-value, single factors on opposite sides of a contradiction.

At the start of this chapter we said, "Words such as 'Contradiction', 'Compromise', 'Trade-offs', and 'Optimization' should set off our problem-detection radar. When we hear any of these words, we need to understand a hidden opportunity is present."

NOTE: We know we have presented a lot of information in the last few chapters and your brain may feel like it is stuffed full. If you feel that way, just skip to the next chapters and come back and read the rest of this another time. You already have plenty to practice. Or you can stick with us and read on.

We've covered contradictions, compromise, and trade-offs. Let's quickly cover optimization because this is another way you can really add value. This is something most engineers and architects and project managers don't even know.

Most people say 'the optimum' is the best you can get from a system. Do you want to know a secret? An optimization point is really just a trade-off. Contradictions, compromises, and trade-offs are often resolved with countermeasures. They take constant vigilance to manage. A solution is a better goal.

To a problem-solver, optimization is a known compromise and an opportunity to deliver better results. To an optimization professional, an optimum is the best you can get from your system. Both are correct depending on how you look at it. But which point-of-view is more useful?

Identifying and resolving contradictions means abandoning the thinking and language of compromise. Trade-offs are our brain's first default process. The brain evolved this way (default thinking) in order to conserve energy, but let's eat an extra bit of food for Brion's[40] sake, and move beyond safety trade-offs!

---

[40] Brion K. Hanks, one of the authors of this book.

There can be no trade-offs with regard to safety. Knowing what we now know, accepting a safety vs. production trade-off is a cop-out. Trade-off thinking is one of the major reasons we are not yet at zero harm within the construction industry. Brion reminds the reader that people continue to be injured or killed in the workplace. Therefore, it's necessary to go beyond this default process, at least for the sake of someone's safety.

So let's end this chapter with a commitment that we will constantly strive for less and less compromise. We will constantly reinforce our pure belief in safety as the way to accomplish better productivity. And then we will consciously address every, "Yes, but..." and deal with every drawback. We want our safe way to be faster than doing the wrong thing and getting away with it. An inventive solution would remove the barriers to adopting a safe way of performing our work.

Safety is serious business and one must be resolute about identifying and eliminating contradictions. It takes energy and a mindset of "No Compromise" to have access to the part of the brain that can solve these problems.

The construction industry's most fundamental safety contradictions have not been fully addressed and there is a lot of work to do. So now we all understand that it is essential to resolve the many contradictions between safety behaviors, need for productivity, need for focus and need for awareness. You need to practice this to get skilled at it. Seek help with your training if you need it.

You who are in the trenches and on the roads and in the cranes and on the 'dozers' see the problems inherent within your specific jobs so it is you who has much to contribute. You have a safety responsibility to yourself and to your co-workers. Therefore, you must think more in terms of contradictions. You must learn to convert complex ones into simple ones in order to resolve contradictions and problems in the workplace. Safety perfection is achievable. But you must take action and solve the problems so that everyone goes home safe each and every day. This is an important tool that helps you thoroughly understand safety.

One More Hint about Solving a Contradiction

Here's one more little hint on how to solve a contradiction. It's a shift in mindset.

In order to solve the contradiction, figure out how to have the best of both sides of the compromise. Being clear about what you want is critical. Then go get it!

Bottom line #7: ***Construction production and safety contradictions are opportunities disguised by your fearful mindset - a belief that you cannot solve tough problems.***

Safety Mantra #7: ***Safety contradictions are opportunities.***

Take-Away Points: ***Chapter 7***

- You are experiencing a contradiction when you have two things that you want but getting one value lessens the chance to get the other.
- Contradictions seem insolvable when important priorities clash. This tempts us to compromise. Construction safety faces many contradictions.
- Solutions to contradictions give you more of what you want and less of what you don't want.
- The common thread between all great innovations is the resolution of contradictions.
- Learn to identify, prioritize, and resolve contradictions as a normal practice on the job.
- Convert complex contradictions into simple ones to make them easier to solve.
- Trade-offs, optimizations, contradictions, and compromises set off our problem-detection radar. These words signal a hidden opportunity.
- The big contradiction in construction is get-it-done-quick versus get-it-done-safe. Resolving this contradiction will improve safety performance and profits.
- Mixed messages between safety and production reveal safety contradictions.
- Our current best safety tools are: constant hammering of safety messages, unrelenting vigilance, and integrity. These are not as reliable as solving contradictions with a safety mindset.

"The status quo sucks."

- George Carlin

"Status quo, you know, is Latin for 'The mess we're in'."

- Ronald Reagan

Chapter 8 – Today's Minimum Safety Practices

We've come a long way since workers built the Grand Coulee Dam during the Great Depression. If this picture was taken today, how many safety violations can you find?

Ottis Rainer Jaquish[41] who had a great personal safety record witnessed men falling to their deaths during construction of the Grand Coulee Dam in Washington State

[41] Ottis Rainer Jaquish was Grandfather to Dayna Hubenthal, one of the authors of this book.

Today there are companies with excellent safety records (but not yet zero harm). These companies use most of the industry's best safety practices. At a minimum, you should be performing these key elements of a safety program:

- Comply with OSHA regulations.
- Persistently identify and eliminate unsafe acts or conditions on the job.
- Build a mentality where all workers contribute to a mutual safety effort.
- Build a culture where everyone accepts responsibility for safety and health.
- Provide employee orientation / fully participate in employee orientation.
- Use the Job Hazard Analysis.
- Have safety meetings every day.
- Train constantly.
- Hold all subcontractors to the same safety requirements.
- Follow up and engage with accident reporting for safety.
- Perform safety inspection tours / keep your eyes open.
- Communicate and display safety goals.
- Hold everyone equally accountable for achieving safety goals.
- Understand the project-specific safety programs.
- Participate and develop a safety recognition program.
- Use the proper protective equipment and tools at all times, for every task.

Personal Protective Equipment

Companies with successful safety programs supply their workers with the correct tools and Personal Protective Equipment (PPE) while on the job. PPE needs to protect the body inside (against absorption and inhalation dangers) and outside (eyes, face, head, extremities, etc.). Also, the work environment needs to be maintained in a sanitary and reliable manner (sort of a large, surrounding PPE).

When employees supply their own PPE, it is the employer's responsibility to make sure it is adequate for the job and maintained correctly. PPE should:

- Be readily assessable to every worker as needed to perform their job.
- Be in good working order.
- Fit each person.
- Be worn by every worker on the jobsite. No one is allowed on the site without the proper PPE. The rules apply equally to all workers.

It is a fact that improved personal protective equipment has helped to eliminate injuries or reduced the severity of them. Great strides have been made when compared to the days of old.

Ottis Rainer Jaquish[42] wearing early safety gear (a "boiled hat") while helping build Grand Coulee Dam in the 1930's

We have little to say about personal protective equipment and that may seem a bit strange since we also are looking at safety through the lens of inventive thinking. Most workers understand what is required and there are plenty of experts to ask if questions arise. Therefore, our major statement regarding PPE is to note that the construction industry does not develop their safety equipment. Instead, manufacturing reps travel to jobsites and witness how their equipment is used. They develop need-definitions from worker interactions. Construction companies do not create the improved PPE design proposals and neither do workers - the people actually wearing the equipment. This seems to be a wasted opportunity.

Remember that we have previously suggested that hidden priorities tell a lot about the momentum within an industry. The fact that workers and employers are not more

[42] Ottis Rainer Jaquish was Grandfather to Dayna Hubenthal, a co-author of this book.

directly involved in the design and creation of PPE is an interesting observation. Hmmm? Does your worksite at least have a suggestion box for improving PPE?

Implementation of Stretch and Flex (Exercising) on the Job

Crews stretch and flex together. It builds team morale, encourages peers to get to know each other and builds relationships. Teams spend about 10 to 15 minutes in group exercise, warming up the body, waking up the brain, getting the kinks out, and bonding.

Stretch and Flex promotes clarity of thought, improves flexibility, and group cohesion. The benefits of this practice are statistically significant.

It is also a good time to impart important safety information.

Kiewit is one of the early adopters and implementers of exercise to improve craft worker safety and focus on the job. It works.

On a daily basis Brion K. Hanks, one of the authors of this book, personally makes it his practice to rotate through the crews on projects where he is the Project Safety Manager in order to participate in "Stretch and Flex" and connect with crew members. This face-time allows Brion to communicate any recent concerns and to recognize individuals for their safety efforts personally. And Brion would have you know that the "Stretch and Flex" has been very beneficial for him personally. The craft workers have expressed the same.

No matter where you work and for whom you work, you can make the time to do stretch and flex and may even convince others to join you.

Employee Orientation

Every time a newly hired employee enters the jobsite, s/he must fill out initial paperwork and watch any required videos, training, or presentations. Orientation should include a drug screening that is done for every new employee, both craft and staff.

Someone (preferably the safety manager, superintendent, or manager) needs to conduct a new-hire safety orientation. This positive first impression is invaluable to set the initial safety tone for the project. During the orientation, PPE is handed out as necessary. Initial expectations are established during the orientation with project priorities reinforcing the safety tone. Some of the important messages to convey are:

- We are serious about safety. We care about you and want you to go home safe every day. We will hold ourselves accountable. And so will you.

- This orientation is taken very seriously and it is made relevant to each worker based on the type of craft person they are (carpenter, laborer, operator, ironworker, etc.).
- Everyone on the project is required to hold themselves accountable and their co-workers as well. This includes individual responsibility to wear the proper PPE at all times.
- All craft workers will be given the correct tools for every job. And they are expected to use them. "If you do not have the right tool, take the time to get it." This needs to be constantly reinforced.
- No one person is the 'policeman' on the project. Each person is empowered to correct any unsafe act or condition.
- All people will abide by company policies: do not take shortcuts even if there is time-pressure to get a job finished.

One of the concepts Brion is able to communicate to the new-hires is their value as a fresh set of eyes. Because they are not yet caught up in "this is the way it's done here" mentality, they can give invaluable feedback. Brion tells all new-hires, "We want you to look around where you are going to work; we want you to ask questions about your work conditions; we want you to speak up."

The manager needs to explain what a Job Hazard Analysis (JHA) is and why it is important. Everyone must review and sign their JHA before starting work.

Brion tells all new-hires, "I am not (nor will I be) a policeman on this project. You are empowered by Kiewit's First Rule of Safety to correct any unsafe act or condition as a part of holding yourself and your co-workers accountable."

After the initial safety orientation, new-hires should attend another orientation; this time with their supervisor. And finally, the new hire should attend a third and final orientation with their foreman. While the new-hire process may take approximately three to four hours, it allows each manager to reinforce policies and procedures through the key people interacting with the new-hire.

The Crafts' Lifeline Is Training

During the year we wrote this book, Brion was on a large project with over 400 people working. There was a great deal of elevated work on walls for which fall protection equipment was utilized. Brion conducted safety training, personal protection equipment training, confined space training, power tool training, rigging and signal person training, and procedural training. On that project two people fell, but as they fell their fall arrest system properly activated. They were able to perform self-rescue by regaining 3-points of contact with the wall and safely climb down with no injury, and best of all they were alive. This is very important safety training.

Whenever you have people performing elevated work, they need to maintain 3-points of contact with the surface. These workers were using proper tie-off methods and are an example of safety success.

After those workers regained 3-points of contact with the surface and, while still using 100% tie-off, they safely climbed down from the wall. No one was hurt because they did everything right: their harnesses fit properly, the harnesses were on correctly; they were tied off above their heads, and they were using shock absorbing lanyards. It's always a good thing when the end result of an incident is that nobody got hurt. Good training and proper use of PPE yields positive results.

One of the biggest PPE hazards (yes, we intended to use the word hazard) is caused by failure to train people on the proper use of PPE such as hard hats, safety vests, safety glasses that fit properly, use of the proper gloves for each operation, and respirators, etc.

Proper training in the use of small tools, powered tools, and fall protection equipment helps to improve worker safety and, through awareness, reduces hazards.

Of course, constantly upgrading people's training on the proper use of heavy equipment is also preventative: designated operator training; confined space training, extensive training to operate cranes, etc. In many cases, it is determined that one or more pieces of heavy equipment, through misuse or failure to follow safety policies and procedures become root cause factors in an accident.

When people are not properly trained there can be (and often are) senseless accidents with deadly consequences. All too often, innocent workers get hurt or killed because of unintentional action of untrained people.

Once a week, there needs to be some type of tool training as a minimum safety practice. For example, it could be about proper use of the circular saws, drill motors, cutting torches, drill presses, grinders, storage of fall arrest equipment, hydraulic wedges, pneumatic nail guns, utility knives, proper use of ladders, and lock-out/tag-out procedures, etc.

Also, when field supervisors themselves are not properly trained, they can unknowingly or unwittingly allow people to do something for which they were not fully trained, and who lack sufficient experience to safely perform an operation. Here's a case in point.

In 2011, Brion heard about a fatality accident on a project forty miles away from his own, and he shared the incident as a lesson to reinforce safety on his project.

That other project was finished and demobilization was down to literally the last vehicle - a water truck - that needed to be driven off the project. The proposed driver did not have a CDL (Commercial Driver's License). The driver also was not experienced, but was allowed by the company's supervisor to remove the truck. The driver did not wear a seat belt. Once under way, while attempting to drive the water truck off the project, the truck veered off a small embankment. The driver was partially ejected through a side window of the truck when it rolled and the worker died. This is a fatality that should never have occurred for so many reasons.

Insufficient training can be deadly especially when you are a frontline supervisor. The fact is that training works. Develop the awareness that trained people will deliver your expected outcomes, i.e., working safely. Failure to respect that training can cost lives or get people hurt.

In the current construction environment, it seems that many do not understand the value of constant training. People are hurt or killed because of this ignorance.

The bottom line: in your safety culture, train everyone and train constantly.

Statistical data shows that unions provide a great deal of training for their members. Fatality accidents are far fewer wherever union work is taking place. Also, in general, union workers are better paid, have better health care and better pension plans as well. Companies that afford better paid employees generally want to train them well and retain them because start-up employee costs are higher than continuing costs for employees.

Large union construction companies usually have strong safety programs which include training for Designated Operators, Fall Protection, Confined Space, Crane Use Compliance, and Power Tool Training. Accident statistics show this makes a positive difference.

Public-funded work such as highways and bridges generally demand trained workers (and often union workers) be used.

The Job Hazard Analysis: a Tool for Building a Culture of Safety

But training alone has not enabled workers or companies to reap 100% safety gains. One of the best safety practices in construction today is the use of the Job Hazard Analysis (JHA) method. Its use in construction creates real synergies between safety and production. In order to put together a JHA, workers and supervisors must be trained for the job they are about to perform, which makes them more efficient at their job. As well, the JHA creation means the workers and supervisors must understand all that could go wrong and how to avoid that. When both the supervisors and workers are jointly involved in the creation of their own safety plan, we see greater buy-in and more adherences to the training.

Early on, Kiewit adopted and evolved the use of this technique. A Job Hazard Analysis program engages every work team. Each crew is required to develop their own JHA where individuals are to be engaged in this process under the guidance of their foreman and superintendent. The goal is for the crew to predict all risks that might come up as a result of doing the tasks of their operation. Then they develop preventative methods and actions.

Brion has personally seen the positive effects of its implementation across Kiewit companies. There is absolutely no doubt that JHAs usage to identify risks and hazards has dramatically reduced accidents and saved lives. When people get hurt (when failure occurs) it's almost always because the JHA was violated.

The JHA is a "living document". It is updated and revised as often as required due to changed conditions or circumstances. If a crew member notices something that is not on the JHA, s/he needs to bring it to the foreman's or superintendent's attention so that the JHA can be modified, reviewed and signed by all.

When crew members sign the JHA, it becomes their own personal safety contract. They are responsible for following this plan and ensuring their own safety. They also commit to watch out for co-workers.

- Every person on the site must sign-off the Job Hazard Analysis that they help create and was reviewed - with them - by their supervisor.
- Signing the JHA means "You understand what you are signing, have witnessed the working situation, agree to abide by the JHA dictates, and authorize its use."
- If any crew or crew member violates their JHA, then they are held accountable depending upon the severity of the violation or incident that occurred. The range goes from: being warned verbally, being written up, suspended without pay, or terminated.

The JHA is now part of the daily conversation and is built into the crew's mindset. Each person must hold both themselves and fellow co-workers accountable, and this becomes an important part of the job going forward.

Developing the Job Hazard Analysis under the guidance of a trained safety leader makes the danger real and present before it occurs. For each step of an operation, three questions are asked of the craft workers performing the job:

1. What's the worst thing that could happen?

    - The crew lists the possible issues and then they list their preventive solutions so the problems do not occur.
    - Team members can give examples of problems and preventions, which varies from operation to operation.

2. How are you most likely to get hurt?

   - The crew lists any possible way they could get hurt and then they list the preventative measures so the problem does not occur.

3. What other risks do you need to be aware of?

   - The crew lists the possible issues and then they list their preventions so the problem does not occur.
   - Sometimes a safety professional will remind team members to look at their environment and other operations that may affect their own work.
     - For example, a welder working on a small frame on the ground might forget to look overhead or might not realize when heavy equipment rolls past, the ground shakes and metal piles could shift.

Using the Job Hazard Analysis is an excellent way to deliver an important contribution to a culture of safety. Additionally, *company policy, management and all leaders must convey that there will be zero tolerance when it comes to violating the JHA, which is a worker's contract to work safe.* In fact, any violation must be immediately and strongly dealt with in a consistent manner that leaves no room for doubt that consequences will be meted out.

Consider how valuable you will be when you are fully engaged in creating your Job Hazard Analysis. Be committed to following it.

Plan Before Each New Operation

Before each new operation, the worker needs to refocus on what they are doing and what the risks might be. The Job Hazard Analysis is a great and formal way to do this before each work operation. But even if your company tells you to "not waste time doing a JHA", you should do an informal one. Lead by example and casually start the conversation before each new work operation. You can even do a mental JHA. Thinking through the work process helps you stay safe.

Developing a JHA presents a challenge, however, because workers need to stop work - yes we said stop work - when there is any changed condition that alters the original plan. Modify the operational Job Hazard Analysis to address the new conditions. Change is real in construction and it needs to be properly dealt with in a manner that always looks to worker safety first.

Goal Setting and Accountability

Each person should have a safety goal that supports the crew's goal and even the company's highest safety aspirations and they should have a way to hold themselves strictly accountable for it. Projects that include continuous improvement of processes with the goal of 100% zero harm indicate a resolute will to improve safety.

The Power of High Standards

One day Brion toured work operations and approached a single red-hat worker (someone on the project less than 30 days) and he asked how things were going? The carpenter indicated he liked working for Kiewit because "Kiewit is serious about safety and they care about their workers. I see it every day."

"That's good to hear. How does that affect you?" Brion asked, expecting a work-related example.

The carpenter then said, "I am so impressed with Kiewit's safety program that I'm using it at home. I have seven children; it's difficult to keep my eye on all of them at the same time. So what I've started doing is we have family safety meetings. Each morning before school I gather them around and I talk about safety, their safety. My children are the reason I work safe each day... they depend upon me."

You can imagine the positive effect that had on Brion. His passion for safety found its way into someone's heart and family life. A safety program is working as intended when it positively changes someone's family.

The workplace is like that carpenter's family with the various personalities and problems coexisting in one place. The workplace is often dangerous and it is impossible for the safety manager to keep an eye on all activities at once. The workers must help when it comes to dealing with their safety and they must also watch-out for their co-workers' safety as well. That is what a family does. It takes care of its own.

Great safety performance begins with great expectations. We must expect more from ourselves and from others to improve and create an environment that supports safety perfection. By communication of high expectations the power of the "Pygmalion Effect"[43] can work its magic regarding safety in the workplace.

Passion, commitment, structure and action are a powerful combination to advance safety mastery. If you have high expectations for yourself, others will see it and be strongly affected.

One way to set high expectations is to highlight the safest crew on a project. Find out their successful practices and talk about them. Use them as a benchmark. Set a goal for the other crews.

---

[43] "The Pygmalion effect", or "Rosenthal effect", is the phenomenon that describes how high expectations (placed upon people - often children, students and employees) help them perform better.

Set Clear Expectations

Currently, the best construction companies create a culture of safety by ensuring that management takes custody of setting clear expectations in the workplace. Management sets goals based on safety first and then sets realistic production goals that don't compromise this. When management uses lip service regarding safety, workers can quickly tell whether or not they are serious about safety. Safety hypocrisy equals accidents happening over and over and over again.

Everyone must hold a bias for human safety. It needs to be absolute. Total acceptance and compliance is expected and demanded. Accountability cannot be escaped by anyone - not today, nor tomorrow. The drumbeat in the workplace must always be ***safety first, safety always!***

One of the highest forms of expectation is, in the moment, having the courage to speak up as a means to make the workplace safer. Expect that you will stand up and ask questions in order to confirm or understand 'the plan'. This is so important, we'll repeat it. Every employee must feel free to ask questions. In fact, they must be expected to do so. It takes communication, communication and more communication.

One approach for supervisors is to include safety reporting, knowledge, and the personal safety record as a part of every employee's review. This demonstrates that leadership is watching and is committed to accountability on every project.

Tell your employees that they should never walk past an unsafe act or condition. They are expected to question any unsafe action they are asked to perform. Listen to your co-workers input if they think you are doing something unsafe. Really consider their input. You can always call the foreman or safety manager for clarification and to help resolve the issue if there is a doubt. Always go with the more conservative (safest) choice in the short term.

When you constantly set clear expectations you leave no room for doubts about safety - no room for compromise.

People to People Communication

As a part of getting to know your people, foremen and supervisors need to talk with employees as a means to determine their experience levels, to learn who the natural leaders are and to communicate that at-risk behavior will not be tolerated. Further, company communications must convey that safety is never, ever to be sacrificed for production or meeting a schedule.

Teaching opportunities are a good time to breakdown workplace barriers. When teaching, remember that in order to get respect you must give respect. One of the best ways of doing this is to solicit feedback from workers. Everyone must listen, validate, and express genuine concern for worker safety. As well, leaders must promote personal responsibility at all levels in the work environment. This is part of building trust that has to be absolute!

One day Brion[44] was touring a project as safety manager when he observed a worker on top of a loaded flat-bed truck, which was at least eight feet in the air. Brion walked over to the truck and calmly said to the individual, "Do you realize where you're standing?" The individual got down on the flat-bed portion of the truck, stuck out his hand to shake Brion's and said, "Thank you for not yelling at me. I was so focused on untangling a strap that I didn't realize it... I was too focused on the job at hand and not focused enough on how to do it safely. It would have been even worse to have been startled into awareness at that moment." This was a long-time employee who generally lived best safety practices but had a lapse - people do weird things sometimes. The respectful means of communicating got the job done without causing more problems and in the future he'll be more aware of his over-focusing tendencies.

It is your responsibility to learn best communication practices. Go out and get that knowledge.

Some companies have a top-down "do as I say and don't ask questions" type of management style. These companies do not explain the reasons behind new safety initiatives. If you work for one of these companies, it is your responsibility to understand the benefits of new safety protocols and then spread the news.

If you are a supervisor, you will get better buy-in if you take the time to explain why new safety practices are required. Better yet, work with crews to identify safety hazards and find safe solutions, then share those among the other crews and up the hierarchical ladder. When you work with teams to solve problems, everyone is more likely to commit to the solutions. Be visible. Have person-to-person conversations about safety. Gather feedback and listen for new issues, too.

A communication breakdown in the work environment is a failure that falls directly at the feet of leaders (whatever their level of authority). Remember, most of today's safety initiatives are countermeasures. That means your safety communications need to be continuous and present. In this way, expectations stand on a safety foundation that is strengthened and core values involving right human relations are reinforced.

It is in this safety arena where you will see safety success. Leaders will be using their safety communication with one main intention: keep people on the right (safe) course.

---

[44] Brion K. Hanks, one of the authors of this book.

Safety Meetings

Before work begins, supervisors and foremen discuss the crews' work during that shift. If necessary, any crew member can share new information or discuss an issue. This is a good time to review any incident alerts in the form of a proactive lesson. This helps workers realize that "it could happen to them" if they are failing to follow best practices and be in compliance with their own hazard analysis.

Every group can discuss their relevant "Play of the Day". Take the time needed so that all crew members understand the tasks for the day. These meeting activities set up each crew for daily success. The safest companies spend millions on these key methods for long term company safety success.

Everyday crews fill out a Daily Snap Shot together because "Safety is no accident." Also daily:

- Foremen ask their crews what is safety relevant to your operation today? This might include: the biggest risk, toughest task, little 'stuff' that could hurt us, basics of their operation.
- As a part of engaging their crews, the foremen also ask, "How can we improve our access? How can we improve our housekeeping? What tools and equipment are we using today that could hurt us?"
- There are reminders such as: is your Job Hazard Analysis current and is the JHA reviewed and signed by each person on the crew?

It is during this time that daily visual inspections are completed for all equipment in use.

It is a good idea to have a monthly Mass Meeting for all project personnel that is led by the safety manager, project manager or superintendent. That meeting allows for everyone to gather together at one location so updates, progress reports, training and craft recognition can take place.

Safety Inspection Tours

After the crew begins work on the project, the safety manager or job superintendent should tour the jobsite for a few hours each day in order to connect with people and look for safety issues. This is a chance to have his presence noticed, build relationships, and observe the reality of work operations.

If a safety issue is observed, the safety manager together with the worker(s) can correct the issue on the spot. Sometimes it's necessary to stop the operation and gather up the crew to discuss an unsafe behavior or condition and reconnect the crew to "safety first". Sometimes it is necessary to have the crew revisit their hazard analysis and modify it because it is a living document in order to make the work operation safer. During this time, the safety manager can and does take photographs in order to share them and later use them for data and historical purposes.

The safety manager should always be available via the radio for any issue that requires his immediate attention and expertise. Constant reinforcement of company policies, procedures, and expectations is important.

It is also imperative that foremen and supervisors periodically step back from their work and look at the larger picture. On occasion, when Brion sees a foreman with tool belt on and head down, he will ask that foreman to take a minute and step back from the work in order to get a better safety perspective. If you can't see the forest for the trees, then you don't know when the forest is in danger. Brion tells all foremen to lift their heads and see what the crew is doing. Look first from the standpoint of safety. Notice what other crews are doing. Is there anything that could cause his/her crew harm? Then make adjustments as needed, and never ignore a changed condition that can potentially injure anyone in the area.

And finally, each worker should perform their own safety inspection before beginning work. Check your PPE. Check your tools (including cords and set-ups). Check your environment. Check out the people around you. Perform an internal check on yourself: are you focused, concentrating, and big-picture oriented? Feeling good? If not, pay attention to what is most important in this moment.

Behavior Based Safety[45] (BBS) Enhances a Culture of Safety

Behavior Based Safety (BBS) is a useful tool that supports a culture of safety improvements. It is a safety program used to find out the root cause of an incident. If you want better safety, you must know why an accident occurred. The problem is people don't want to report their foolish mistakes. BBS strives to change this through cultural rules focusing on behavior and not blame.

A good BBS slogan might be, "Don't fix the blame, fix the problem!"

---

[45] Herbert William Heinrich worked for Traveler's Insurance Company in the 1930's. He noticed accident patterns as he studied many accident reports. The most obvious pattern was that nearly all accidents were due to human error. He began the work of Behavior Based Safety, but the actual name was created by Dr. E. Scott Geller of Safety Performance Solutions (SPS) in 1979.

We will comment on human error and safety later in the book. Innovation Scientific knows that high incidences of "human error" as a root cause in safety issues is an indicator of a systemic misunderstanding of safety and quality as a process.

Since BBS does not embrace resolving contradictions, it remains a form of compromise. We call on BBS leaders to teach resolving contradictions as part of their program and leverage the language to enhance reporting of accidents and frame the challenges.

Everyone in construction knows that some people indulge in flawed thinking patterns so best safety practices are put in place to mitigate the unproductive thoughts expressing into action. We must also discover why people act on their foolish thoughts.

A supervisor observes a work activity and then gives real and immediate feedback - reinforcing good observed behavior as well as correcting unsafe behaviors. When performed consistently, safety will improve in any work environment.

BBS is one answer to that challenge. Recent important safety gains have been realized by using BBS and it has been shown to be an excellent best practice in safety. Use this sincerely with constant and principled dedication and you will see safety improve. We are not going into great detail here, merely introducing you to some of the basics. There are books on the subject and you could either educate yourself by book or take a class and earn a certificate.

BBS is predicated on the fact that people make errors and compromise when they know better. This safety tool allows work groups to identify, to measure and to change the lead-up actions (or activators) that initiate unwanted behaviors and produce negative consequences. A successful BBS program discourages repetition of these activators.

BBS wants the company goals and expectations to be consistent as well as delivering clear constant communication. Behaviors and consequences are observed, all data is collected, analyzed, and the feedback informs future decisions. Feedback is given and immediate correction(s) made at the time of the BBS observation. It is important to understand that the purpose of all this data collection and analysis is to fix the behavior problems, not assign blame. It's more important that everyone knows what happened and why it happened, but not so important who did it.

For BBS to work, data collection must follow certain criteria. Descriptions must be SMART - **S**pecific, **M**easurable, **A**ttainable, **R**elevant, and **T**ime-bound.

Also, BBS is supposed to include all aspects of the business - no one is above the law - no department is exempt - safety behaviors will be defined for everyone. In fact, BBS process maps are created for each operation or process and all critical behaviors are identified to improve safety performance. There are other controls, audits, and actions used, but we will not go into detail here. This section is not meant to be an inclusive education on BBS; rather, it is supposed to introduce you to a very good tool for safety today.

Adoption of this method is growing and is perceived by many to be a cutting-edge practice. This is one of the areas the supervisor or worker could gain a competitive advantage by getting certified in BBS or at least taking a class or reading a book.

Red Diamond Event Tool Enhances a Culture of Safety

Red Diamond Event, developed by Exxon[46], is another tool that supports a culture of safety improvement.

Focus on the Diamond

In 2010, Exxon looked at past company history to learn from accidents or near misses. They took the typical Safety Triangle, which has:

- Fatalities at the top;
- Lost Time Incidents one step lower;
- Recordable Incidents one step lower;
- First Aids and Near Misses are at the bottom of the triangle.

This creates a red diamond within this triangle as a more "refined paradigm".

At this point, a "what if" question is asked, "If you run the scenario of an accident (or near miss) 100 times, would it result in a fatality?" Through investigation and getting to the root cause, if the answer is "yes", then it is disclosed company-wide from a standpoint of lesson-learned as well as how to avoid it in the future. Exxon does not want to repeat accidents!

After a near miss, have you ever heard any of the following phrases: "If not for luck..." or "We got lucky this time..." or "We dodged a bullet that time..."? If so, then you are at risk for a fatality tomorrow. No one wants that.

---

[46] ExxonMobil is the world's largest publicly traded international oil and gas company built on a concept of global businesses. They operate facilities or market products in most of the world's countries and explore for oil and natural gas on six continents - crude oil, natural gas and petroleum products. ExxonMobil employs about 76,900 workers. According to Rex W. Tillerson Chairman and Chief Executive Officer, in the *Operations Integrity Management System* report, ExxonMobil remains steadfast in its commitment to excellence in Safety and believes that all safety, health and environmental incidents are preventable. ExxonMobil intends to promote and maintain a work environment in which each of us accepts personal responsibility for our own safety and that of our colleagues, and in which everyone actively intervenes to ensure the safety, security and wellness of others. We believe these measures will help drive us closer to our vision of a workplace where "Security is Everybody's Business," and where our environmental performance meets our expectations to "Protect tomorrow, today."

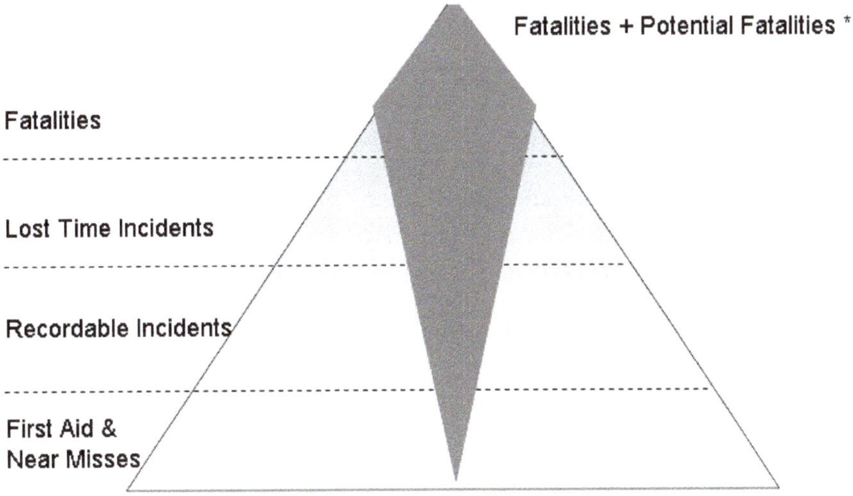

The most valuable data for preventing fatalities and serious injuries is hidden within the diamond!

** the only difference between a fatality and a potential fatality is luck (or lack of luck)*

Safety Triangle

Focus on the Diamond is becoming an important part of the industry's safety program. It is producing positive results. Once again, it might be of benefit to you to learn more about this subject.

Project-Specific Programs

Brion studies industry best practices for safety and strongly believes that Kiewit is one of the leading companies for project-specific programs. Their programs begin with an in-depth new-hire orientation process that is vital for setting the safety tone right out of the gate. A safety orientation conveys company policies, best practices, project-specific top hazards, and is important from a first impression standpoint. Literally hundreds and hundreds of Kiewit employees have taken the time to tell Brion that Kiewit is the "safest company" where they have worked.

Kiewit also gives Weekly Toolbox Training on how to use different types of equipment, such as power tools. They give training modules for all foremen such as New Hire Orientation, Effective Communication, Hazard Recognition, Eliminating Unsafe Behavior, Holding Your People Accountable, and Crew Engagement. Kiewit requires and performs Designated Operator Training for cranes, forklifts and Bob Cat (Skid Steer). They have trainings for Rigging and Signal Persons. And, as needed, Kiewit trains its employees for trenching and confined space work.

In addition, all workers are trained to know that every Kiewit operation is required to conduct and maintain a Job Hazard Analysis. It is signed off by the crew before beginning the operation and it is to be modified when any changed condition occurs.

Kiewit also has a great company-wide program called Speak Up / Listen Up where people are encouraged to speak up about safety first and any concerns they might have. Supervisors and management are then trained to listen and act on whatever has been conveyed to them by employees.

How much of this does your company offer?

Subcontractor Safety Management

Subcontractors may or may not have a special interest in safety, but subcontracting is the worker group that must take the most responsibility for their safety. Having said this, they are usually the workers with the fewest resources for safety, are usually the least trained in safety processes, and are usually considered by others to be the most expendable on the job.

Subcontractor safety is often ignored by the prime contractor who expects them to 'handle it' themselves. Employee crews often resent subcontractors and are not as watchful for their safety. And subcontractors often must provide their own tools and PPE, which may be inferior to the employee crews. They are usually less trained and they usually do not have a consistent set of safety rules and practices to follow because each project and each boss probably sets different safety standards. Because of all this, they have weaker buy-in to safety expectations and are more likely to have an accident.

Over the years Brion has observed subcontractor employees doing unsafe things which violate OSHA Standards. When he brings it to their attention, they regularly are not aware and were never properly trained in safety. This reason alone accounts for many people being injured or killed on the job.

When it comes to project safety, you will not have the safety success you need without educating all of your workers on site. You must set clear expectations and hold everyone on the project equally accountable. From a safety standpoint, subcontractor employees must be managed the same as prime contractor employees. There can be no difference between how people are treated, managed and recognized for safety efforts.

Accident Reporting for Safety

Over the course of his construction career, Brion K. Hanks, one of the authors of this book, has seen the whole range of accident reporting. He's heard about companies that fire workers for reporting issues. And at the other end of the scale is

Kiewit who has a policy to report ALL incidents. It does not matter whether an employee gets a bump, a scratch, a bruise or minor cut; it is made abundantly clear to all employees that they are to report an incident, no matter how minor, immediately to their supervisor. Should first aid or further medical attention be required, Brion and project management become immediately involved.

Why do safer construction companies require this level of reporting when it means a lot more paperwork is required? Kiewit discovered that reporting every minor incident led to reduced accidents. For example, a minor incident is reported and investigated. It is determined a nearby hazard existed due to poor housekeeping. Thus, the problem was fixed and also discussed at subsequent meetings. No other accidents occurred where housekeeping was at the root. Since initiating this program, Brion has seen a dramatic reduction in the number of recordable accidents or incidents.

Safety Recognition Program

Brion has noticed that recognizing solid safety practices yields great safety gains. A heart-felt handshake, acknowledgement of safe work, and noticing solid safety choices, makes workers feel great and motivated to continue that trend.

Sometimes, Brion likes to do even more. For example, he has noticed that people like to wear the company's logo gear: hats, shirts, sweatshirts, and coats can all be given as recognition of good safety practices.

Other coveted safety recognition awards are project-specific coolers, various types of carrying bags and tools. Workers especially like it when the year or project name is on the shirt. For example: because they may have twelve shirts - all from different years – they can show how they have been an outstanding safety worker for a long time.

Workers can be recognized for their safety efforts because they corrected an unsafe act or condition, made suggestions for how a work operation could be performed more safely, or identified an issue or new hazard.

As well, people are recognized for speaking up on any matter that improves project safety. Brion has seen great pride exhibited when individuals are recognized for their safety efforts. Brion also makes it a point to do shout-outs to people as he tours operations where he personally thanks people for being tied off when doing elevated work, for proper use of ladders, for chalking their tires when getting out of large vehicles, for grounding their implements when they finish a task or when workers use three points of contact when getting on and off equipment.

Any and all methods that reinforce safe behavior will be effective in reducing incidents or accidents on a project. Again, the object should always be safety perfection, i.e., everyone going home safe each day.

The Construction Industry Safety Leader[47]

As we concluded the research phase of this book, there was one company that appears to match our vision of systematic safety as a science. After an industry review, the authors determined that this company is *the* safety leader in construction. The company is Bechtel Group Inc.[48] Admittedly we do not have an inside view of the company's total practices but the outside view leaves some important clues to the trained eye.

Why do we conclude that Bechtel Group Inc. is *the* safety leader in construction? First they have been dedicated to the idea that all accidents are preventable for about 20 years.

The company has committed to "achieving and sustaining zero-accident performance through continuous improvement" so Bechtel leadership understands that the zero-accident performance must be backed with a credible approach.

Among Bechtel Group's 47,000 employees, about 1,000 are Environment, Health and Safety (EHS) professionals. These employees are distributed at all levels of the organization from boots on the ground to the top levels of management. Their project planning process incorporates ESH systems and is one of the final considerations of whether a project is a "go" or "no go"

In fact, about 20 years ago, Bechtel adopted a systems approach to ESH which moved them into prevention mode (before problems occur) rather than a reaction mode (dealing with problems after the fact). This is very important step into the future of construction safety. If they understand the value of prevention and act on the goal, it is not just talk.

---

[47] Information generated by Innovation Scientific's indicators of process competence.

[48] Bechtel is a privately owned engineering, construction, and project management company, which is headquartered in San Francisco, California. They have offices around the world and more than 53,000 employees. In 2012, their revenues topped $37.9 billion U.S. dollars and new contract awards valued at $23.9 billion. Their diverse worldwide portfolio includes: energy, transportation, communications, mining, oil and gas, and government services.

Also, information garnered from the article written by Sandy Smith, *Bechtel: Striving to be the Best*; 12september2003; http://ehstoday.com/safety/best-practices/ehs_imp_36578 (last accessed on 07september2013).

Over the years Bechtel has integrated safety into business processes. They have enough processes that are "stable and in control" that they can use Six Sigma with their continuous improvement programs. This is an important step! Their next steps are to make safety systematic and more scientific.

In 1994-95 they began to incorporate BBS into the culture - in fact they have evolved it beyond most people's understanding. Also, subcontractors are required to live up to the same safety requirements as Bechtel employees. Bechtel has introduced new PPE to communities in which they work where they are the observable safety leaders and change agents. In many of these places, safety was not valued highly. Where ever they work, they train their people at all levels and constantly - and they are using technology to create fun, interactive simulation-based safety games that are as effective, or more effective, than in-person trainings. In fact, as far as we can tell, they perform most of the industry best practices listed in this book and have begun the process of performing some of the new best practices that we are suggesting.

Bechtel also uses every successful communication strategy the industry has identified to improve safety: weekly safety meetings, JHA, high performance teams, distributed monthly safety statistics, and Bechtel Briefs. In some locations, Bechtel computers boot up with weekly safety messages. They have a wellness letter that includes information for families and they extend their communication efforts to the local communities as well.

One of their common quotes we particularly admire is:

> *Time and again our work has demonstrated that the only limits on human achievement are those that we place on ourselves. - Unknown*

On their website, Bechtel says that their three main values are integrity, quality, and zero accidents. Their projects help grow local economies and improve the quality of life for communities and people around the world. Their culture is grounded in integrity and respect and the highest standards of ethical business culture and their reputation for adhering to these standards is one of their most valuable assets. *[Remember the reputation bank account? Bechtel seems to understand this basic concept.]*

In our opinion, this company has a good shot to be the first construction company to achieve zero accidents but we also think that is not likely to happen until they adopt a higher goal of 100% zero harm. The key to future success is to set much higher expectations in order to force even deeper understanding and professionalism. In the future we see the industry leaders will claim:

> *"Processes and projects will have safety and quality built-in, with cost and performance as prime considerations and meaningful work for all."*

Bechtel is well on the way to achieving many of the goals in this book. They also cherish the most important core safety values. And they seem to be a living example that safety does NOT get in the way of profits. We cannot determine from our outside view, however, if they realize that safety is the path to greater profitability and competitiveness. Since they are one of the industry leaders in both profits and safety, we think they might realize how to leverage this synergy.

If you choose to model one industry safety leader consider what Bechtel is doing. The future may be coming faster than we imagined when we began this project!

Involvement at All Levels

The current situation from OSHA's and business points-of-view: workers rely on companies to manage safety in the workplace. OSHA sets minimum safety standards and regulates compliance. The industry relies on outside manufacturers to develop improvements in equipment, etc. But, the construction company has final control and, therefore, final accountability for safety in all work environments.

The current situation from the worker's point-of-view is the same as above with one addition: each individual has full control of their decisions and choices and, therefore, personal accountability for safety.

- Everyone must create cultural momentum for "safety first, safety always" without compromise.
- Everyone understands the clear expectations for safety in the workplace; everyone follows the rules.
- Everyone holds each person accountable independent of their level of delegated authority.
- Everyone follows-up and contributes on identified safety issues. Regarding safety consequences (both positive and negative) and cooperation is pervasive.
- Everyone understands the expectations regarding safety.
- No one makes excuses or fails to follow the rules.

No longer will it be acceptable to rest on your company's laurels when people are still hurt or killed, even if it happens at a lower rate. The best people in the industry believe that safety improvement will come about by creating better safety cultures that involve more discipline and individual accountability from the highest to the lowest levels of the work force. Every safety culture needs leadership, which means that you will work tirelessly for the greatest good - which is zero accidents or, even better, 100% zero harm. If you follow our advice you will also optimize profits as well.

Some may stubbornly argue that no more can be accomplished. But a leader knows if we do not first step onto the path, we can never reach our zero harm goals.

Bottom line #8: ***If your best practices are achieving 95% safety on the job then your priorities are 95% correct. Remember that when you stake someone's life on a countermeasure you must have 100% vigilance and multiple backup plans. To get that last 5% that your priorities demand, you must move beyond compromise and countermeasures.***

Safety Mantra #8: ***Your safety record reflects your integrity and your knowledge.***

Take-Away Points: ***Chapter 8***

- Today's minimum safety practices handle 95% of most safety issues. The last 5% are the tough ones to resolve. This book intends to help you resolve those last issues.
- Set high safety standards and clear expectations.
- Build a team mentality for your safety culture.
- Provide employee orientation / fully participate in employee orientation.
- Job Hazard Analysis, behavior based safety, and "Focus on the Diamond" are excellent best practices.
- Have safety meetings every day.
- Train constantly; it saves lives and makes your safety culture stronger.
- Exercise improves worker attention and conditioning.
- Hold subcontractors to the same safety requirements as employees.
- Follow up with (and engage with) accident reporting for safety.
- Perform safety inspection tours / keep your eyes open.
- Communicate and display safety goals; hold everyone accountable for achieving them.
- Understand project-specific safety programs.
- Participate and develop a safety recognition program.
- Use the proper protective equipment and tools at all times, for every task.

"Safety means first aid to the uninjured."

- Author Unknown

"Learning is not compulsory... neither is survival."

- Dr. W. Edwards Deming

Chapter 9 – The Current Culture of Safety

A project's safety culture is built moment by moment with each decision and each action that is taken. As the project progresses, momentum builds towards more safety or less safety. You, as a worker or field supervisor, need to know what behaviors you must demonstrate to increase safety and how to respond in dangerous situations.

Small Businesses Can Build a Safety Culture, Too

Sadly many small businesses accept defeat before they begin. They believe safety is too expensive and they cannot afford it and remain competitive. This is simply not true.

Eric Jacobson, a small business paving contractor, ran his company for thirty years with only two accidents. One was proven to be a false claim, the other accident involved a walking worker who tripped over his own foot and twisted his ankle. Eric used dump trucks, compactors, excavators, trucks, trailers, other huge equipment, and heavy supplies. He and his crew often worked on dangerous sites - steep hills, under the threat of falling trees, around swollen creeks and saturated ground, in earthquake country, around electric lines, with slippery leaves and slugs underfoot. And after thirty years, only one person sprained his ankle on the job.

When we asked Eric the secret to his impressive safety record, he said he had safety meetings every day on every project. And when they switched to another part of the job, they had another safety meeting. When conditions changed they revisited their safety plan. He always stressed, every day, "Slow down. Think about what you are doing and how you are going to do it. Make sure you are working the safest way. It's not going to save any money to hurry up and get hurt. Slow down and think." He did not tolerate people who could not or would not fit with his culture of safety.

After some in-depth discussions, we thanked him for his information and he begun to leave. Suddenly Eric turned around and said, "Wait, I have another secret."

He also repeatedly told his workers, "I write the checks, but I'm not the boss on the job. We need to figure out how to get this done together and do it safely. If someone else has a better idea on how to get this finished safely, we'll do that."

Eric had extremely loyal crews. They knew he cared about them, knew he took the work and their safety seriously, and they knew he would not put his ego or finances above their good. He involved them in every bit of the work which achieved employee buy-in and commitment to work safely. They were a team and had mutual respect. He ran a successful and safe business.

Eric Jacobson is a shining example of great management in construction - a stellar safety leader the likes of whom we desperately need throughout the world of construction. His performance defies the excuse that small companies do not have the resources to be safe and make a profit. This is possible today.

Entrenched Acceptance of Danger - The Opposite of Safety Responsibility

Although Eric was successful on every level, other companies are less successful. Eric Jacobson accepted responsibility for his crews. Some do not.

Be a part of the solution not a part of the problem. In the current situation, we find that many construction people are stubborn, hard-headed and not fully committed to safety. But perhaps that is only because they cannot see a different way.

From an inventive view, the problem is that real-world construction must address both safety *AND* production. We, as a society, want the benefits of both. This is a contradiction that is true no matter how badly we want to say safety is more important. We continue to experience entrenched, pernicious, unpredictable, unsafe events because we currently do not know how to resolve this contradiction.

NOTE: We realize that the serious reader might be getting frustrated at this point. We assume you are thinking, "Okay, you are right. I see it. I see the contradiction. Now what do I do about it?" This frustration is a good sign that things are happening in your brain and with your emotions. It means your subconscious is already working on this problem for you. This is part of the process. The frustration is actually important and now suggests that you are no longer making excuses. Innovators experience this process repeatedly. You are getting there.

But, hang-on, we are not finished yet with the current points of view. It is important that you realize the full current landscape before you come up with solutions that will work for you.

Rework Affects Safety Responsibilities

Companies want to avoid rework for several reasons. The two most important reasons are increased risk and costs.

- Interrupted schedules or delays can add pressure, which often contribute to incidents, near misses, and or accidents.
- Increased tear-out costs encourages workers to attempt to "make up" those costs by working faster.
- Tear out and rework processes introduce new risks; added to the original danger, risk is at least doubled.
- It costs to perform the work the first time, costs to tear it out, and costs to rework it.

To avoid mistakes, workers must really understand what they are building ahead of time and understand the inherent risks of performing that work. Detailed work plans provided to workers upfront help them follow quality and safety processes more easily.

Putting hold points in place allows workers to check the work and get focused on performing the job safely before anything is "set in stone". Supervisors need to make sure these "hold points" are added into the work schedule. Workers need to take a moment and center themselves on the job. Taking a minute now and again to refocus on the plan is a good practice.

Getting it correct the first time requires that people ask questions if they have any doubt, regardless of how minor it may seem. And if there are any safety concerns, those concerns must be voiced and handled so that the work proceeds with no incident. In this way, the company will reap a significant return on investment - getting it done correctly the first time is safer and cheaper!

Best Practices Everyone Can Afford to Enforce

Frontline supervisors need to set up the conditions to create worker buy-in and good morale. But every leader at all levels within the company is responsible for tracking and ensuring that these goals are met - that means you at what ever position you hold. The following actions help promote positive conditions and are fundamentally important to successful safety programs:

- Ask if proper equipment and tools are provided. Always use them.
- Constantly train yourself.
- Use qualified people on work projects; if you must hire less qualified people, build time into the schedule for training and mentorship.
- Develop detailed work plans that are effectively communicated to crews in the field.
- Make sure there are no language barriers that can hinder clear communication.

- Encourage workers to speak up and validate that they were heard (meaning frontline supervisors are listening and acting on what they are told).
- Stress the need for good housekeeping, and
- Maintain clear access to work areas.

Like spokes on the safety wheel, any one that is broken can contribute to or cause a catastrophic accident. The safety wheel must be greased by management, leaders at any level, and workers in order to allow workplace safety to roll along smoothly and with certainty that it is not going to breakdown. In this way, it resonates throughout the workplace that safety is always first and foremost.

Follow-Through

Once you have trained workers, trained supervisors, and have committed buy-in from everyone, the next important step is to have strong follow-through in the field. This is where accountability on the part of every worker comes into play. It means workers must constantly hold each other accountable because everyone's daily goal needs to be that they go home safe each and every day. That goal is reinforced constantly by action and effort to keep it at the forefront.

Formal and Thorough Investigations Influences Safety

Although it is management's responsibility to conduct formal, rigorous and thorough investigations when an incident or accident does happen, it is everyone's responsibility to participate fully. Sufficient questions must be asked and answered in order to get to the real root cause of any incident, accident or near hit. Investigators and workers alike must be committed to arrive at the root cause and must find the truth (let the chips fall where they may).

In a company with integrity, workers must show the same kind of integrity and dedication to exposing the root causes. It is essential that workers understand "the truth matters most" and that everyone within the company wants that truth so they can respond appropriately. No one wants a similar incident or accident ever happening again. The "fix" has to get to the real cause in order to be effective from the standpoint of avoiding a similar incident or accident.

This is a best case scenario where both the company and the workers strive to put safety first - they are proactive and keep the most important goal (safety now and in the future) in mind. But what does a worker do if the company has less integrity and more "get 'er done at any cost" attitude? Is the company just looking for a scapegoat as they ask root cause questions? Will the worker lose his/her job? The reality is that these fears are exactly the reason people cover up and doctor the truth. The authors wish you will take a moment and realize where these kinds of mindsets and actions will lead.

The time is quickly coming to an end where this kind of irresponsibility will be tolerated. The situation is going to be changing within the construction industry and each worker is going to need to be responsible for their own and other worker's safety. In fact, a lot of the burden of responsibility will be built into the system. In the future, all of your actions on the construction site will be observed and documented - like "Big Brother".

OSHA and other safety regulatory agencies[49] often identify company management failures to be at the root of an accident. In those cases, fines are levied. When a company produces many accidents or the safety problems are entrenched or pervasive, then it becomes obvious that fines are not enough to dissuade this irresponsible company.

But what does an entrenched or pervasive unsafe environment mean to workers? Companies in willful disregard of OSHA rules will continue to have accidents. No fine can restore a person to wholeness or bring them back to life if they are killed in the workplace. This is not the type of company you want to work for. And this is not the kind of example by which you should set your own safety standards. In the current environment, if you work for a company like this, workers have two options. Either quit and find work at another job with more integrity or be a leader and stand up for safety changes. This is a scary suggestion. But personal responsibility is forged in times of stress, not when things are going smoothly. Standing up for safety when there is something to lose is exactly when you build your safety character and commitment.

That's why one of the first things we wrote in the book was "character matters" when safety is involved.

Safety is not static. In fact, safety programs need to adjust constantly to changing conditions and people must be proactive in pursuit of safety perfection. Safety is not just the responsibility of management or business owners or safety professionals. On-site safety is every person's responsibility no matter what positional authority they are granted.

Better Reporting Improves a Safety Culture

Brion has found that the reporting of even minor incidents or injuries takes place at a much higher rate with union work and larger construction companies. Workers are more honest, responsible, and prevention-oriented when they are instructed by management of the mandatory requirement that all incidents or accidents are to be

[49] For example, Mine Safety and Health Administration (MSHA)

reported, no matter how small. This allows management to investigate and make "minor corrections" that then can avoid more serious accidents. Brion has been part of every stage of this process: encouraging the reporting and handling the investigation of all types of incidents. Over time, there has been a direct correlation between reporting small incidents and the dramatic reduction of recordable accidents.

When you put a policy in place that requires workers to report every bump, bruise, scratch, strain, sprain, cut or even getting something in their eye, you are going to see an increase in the number of reported incidents. This provides the safety supervisors with valuable information about the reality of their construction site. What quickly becomes apparent is that these incidents create opportunity to make minor corrections in operations and sometimes tool usage. These incidents present opportunities to talk with the workers and foremen, which in turn builds trust in the relationships and supports the transfer of safety information. It also creates the opportunity to improve and update the JHAs going forward and confirms individual responsibility to not violate it.

On Brion's[50] projects, as long as workers follow their authorized JHA during their operations there are no negative job consequences even if something bad happens. This creates trust in the culture of safety, drives the cost of accidents down, and the daily safety consequence is more workers go home safe. And there is absolutely no better safety consequence than that.

The evidence is conclusive: better reporting leads to fewer fatalities and injuries.

However, not all construction sites are run that way. Brion has heard that many companies fire the worker who reports minor injuries. When these are the consequences of reporting, workers fear a job loss and will often cover up the cause of accidents so it does not reflect badly on them. And, even worse, some people cover up the actual accident itself.

Suppressing the truth is a direct precursor to more accidents. So your choice at each of these decision points is important but may be scary. Lose my job or improve safety in the long run. Lose my ability to provide for my family or condone and support management failures so others will be injured, too? Whenever management fails to be serious about safety compliance, accidents happen at a much higher rate and are often serious or even fatal. Period! This is supported by government data.

So what are we telling you to do at these decision points? Are we telling you to choose integrity and the good of future workers over monetary gains and your

---

50 Brion K. Hanks, one of the authors of this book.

family's best interests? No. And we are not telling you to choose the other option either. This is the choice that exists in the current situation. We are saying, find the third choice!

When workers are fearful for their jobs and cloud the issue, it takes a dedicated effort to get to the truth. Tragically, people get hurt or are killed on the job when misdirection rules on projects. We must be aware of the consequences of unproductive fears in the workplace and drive fear out of the organization.

Pain and suffering are real. Only when you deal with it day-in and day-out can you fully appreciate these facts. Once appreciated, it seems a certainty that everyone would care enough to do everything they can to avoid future pain and suffering. But the truth is money and self-interest rule far too often. And everyone seems to think that these are the only two choices.

Start Now, Because Sometimes Countermeasures Take Time

We were speaking about safety with Ophir Orr, a professional engineer and a leader in the American Society of Mechanical Engineers. He has been responsible for making changes in traditional industries similar to construction. "Sometimes," he says, "it takes years to change the current paradigm. We have to re-tool and re-educate everyone and there is going to be resistance."

"One of the tools of change management is 'the elapse of time'. Seat belts are a good analogy." He continued. "Seat belts usage started while I was a young man and I still occasionally catch myself thinking, 'Oh darn, now I have to put the belt on; and now I have to take the belt off.' But my kids don't even think about it. They get into the car and put on the seatbelt without conscious thought because they have never known a different way. If we take the steps laid out in this book, in 40 years when the new generation is working in construction, they will not even know a different way of working. And the worksite will be much safer."

Because the new generations will have a higher degree of safety built into their mindsets and psyches, these ideas are going to be less of a countermeasure. That's why we need to begin now with unflagging determination to indoctrinate the young workers.

When this happens, the young generation of construction workers will have less of the safety versus monetary contradiction built into their paradigm. They will have more open minds to new possibilities and the safety advances will accelerate along with productivity. More and more often it will come up that social responsibility (like safety) is just as valuable even if it is more difficult to measure than monetary benefits. We are already seeing the seeds of this in today's youth.

We can all see the benefit of formulating short-term goals and implementing those plans (like planting seeds) now for future safety, but make no mistake, we are not advocating gradualism.

> *"This is no time to engage in the luxury of cooling off or to take the tranquilizing drug of gradualism." - Dr. Martin Luther King, Jr.*

There are those with entrenched mindsets who would say these changes the authors espouse need to be achieved in small, next-step increments so we do not interrupt business competitiveness or endanger market share. This is fear speaking and we authors do not agree at all. In fact, the small, next-step method of change is usually a way for resistors to put off doing anything about resolving the problems.

The amazing thing about the innovation process — resolving contradictions, dealing with stagnant mindset inertia, not compromising, and win-win solutions — is that quick, abrupt reforms can be achieved and easily accepted into the minds and hearts of the construction community if performed correctly.

When you have intractable contradictions built from hidden priorities based on fear and pain, then you do not use gradualism as a method for overcoming it. Martin Luther King Jr. knew that would never work as a way to overcome segregation. We know that gradualism will not work to eliminate the safety versus profits contradiction.

Bottom line #9: ***Build a culture of action based on safety priorities. Use best practices but go further. Continuous safety improvement is mandatory for survival.***

Safety Mantra #9: ***A safety culture is made moment-by-moment and choice-by-choice.***

Take-Away Points: ***Chapter 9***

- Business size does not matter. Any company can build a culture of safety.
- There is no silver bullet to fix all of your safety problems.
- Take action on safety priorities and learn. Continuously improve safety through a process-focus and problem-solving.
- Rework increases costs and risk. Do it right the first time.
- Make sure there are no language barriers to communication.
- Encourage workers to speak up. Frontline supervisors listen and follow up.

- Integrity is forged in times of stress. Promoting safety when there is something to lose is exactly when you build your character and commitment.
- It can take a whole generation to embed a new mindset into an old culture so be in this for the long haul.
- Safety is not static. Constantly adjust safety programs as needed.

"To know what is right and not do it is the worst cowardice.

- Confucius (K'ung Fu-tzu), Chinese sage (551-479 B.C.)

"Live as if you were to die tomorrow. Learn as if you were to live forever."

- Mahatma Gandhi

Chapter 10 – Responsibility and Consequences

Take the time to realize that you are responsible for the consequences you create. Every action or inaction produces a cascade effect. This chapter looks at some of the consequences for accepting responsibility or not.

Upon accepting responsibility comes a new awareness of the depth of the problems and challenges. You are going to see things that you did not previously see. You are going to understand things from a different point-of-view.

Part of the nature of construction work today is that it is inherently dangerous. Danger increases when projects are rushed or under bid. Weather can increase risk. Danger is multiplied on cutting-edge projects or on a type of work that has never been performed before. New types of equipment can increase risk, but so does old equipment that is not maintained or used properly. Inexperienced workers raise the risk factor. So can tired workers or…well, there are a whole slew of reasons why people in the mix increase danger during a project. The bottom line is that today construction is inherently a dangerous business. But you have choices.

In this current competitive environment, supervisors and construction workers must understand that when your teams produce an undesirable consequence, it is both a red flag and a true opportunity.

Electrical power, heavy equipment, and power tools in proximity to people can be a prescription for disaster. Gravity affects many outcomes. Dangers lurk in the air and under feet; and lots of different activities happening at the same time can be chaotic and dangerous. Noise distractions, schedule expectations, and egos all add to the risk profile on construction sites. Sometimes negative emotion come into play. In some ways, it's hard to believe that any project can be completed safely.

You will see co-workers endangering themselves and others and they may feel indignant when you call them out. Your mindset and beliefs will help you stand firm when you remember the consequences if you don't.

The more work you do, it would seem, the more opportunities exist for some form of calamity. Will construction always be a gamble? Is the danger unavoidable?

Consequences of Previous Experience

"If you only have a hammer, you tend to see every problem as a nail." - Abraham Maslow

If you never look at the construction site differently, you will not see new safety possibilities. Do not fall prey to the number one trap - using only what you currently know. Here is an important skill you must learn if you are to remain relevant as the industry changes: you must learn excellent problem-solving skills. Part of that skill is developing the flexibility to look at things from different points of view and listening to different points of view.

Without this you will rely on your existing toolset, which means you will continue to produce the same results as you have in the past. You will perform your work as you have always performed it. But as the industry starts making these changes to embrace safety more systematically, scientifically, and process-oriented, your old ways of working will degrade in effectiveness.

As other supervisors build safety into their crews, you will have more accidents, probably start punishing errors more harshly and must constantly increase your vigilance. It will cost you more to run your crews than other people. As the industry changes, you will become slightly stagnant. You will be a less visible safety example and will need to require more safety inspections (countermeasures).

We are not saying those skills are not important. We are saying that you will need much more than that to succeed in the long run. When you reach a new level of awareness you must embrace new tools to match your new awareness. And one of the most profound tools is the ability to identify better problems to solve.

But it is important to direct any new-found "evangelism" toward you first and start growing yourself. Be the example first. Once you are a true example, you can begin to introduce your new knowledge to other people with integrity and lead by example.

Your first mission (it's hard work) is to identify all countermeasures versus solutions to problems within your area of expertise. Notice when you jump to a countermeasure. Pay attention to the effectiveness of your actions. Pay attention to when you got lucky rather than were actually safe. Learn to acknowledge when you compromised on your desired outcomes. Expand your perspective and your problem-solving skills.

Consequences of Mistakes and Unproductive Job Fears

A mistake has been defined as a blunder, misstep, gaffe or lapse. Call it what you will, mistakes can lead to unexpected and undesired outcomes. Even small mistakes can have big consequences. So they should not be ignored.

Mistakes are also an important part of improving safety. The consequences of those mistakes can be hard to accept or they can be viewed as an "unexpected learning experience". It takes humility to acknowledge that there is more to learn even when your knowledge is world-class.

How do we channel "unexpected learning experiences" into everyday construction life? One way is to study OSHA, MSHA and company incident alerts with the intent of learning why "accidents" or incidents happened. Gaining an understanding of the root cause(s) is vital to moving forward in a positive way with the goal of prevention. Of course, accountability must be understood and consequences meted out and corrections need to be made as a means to prevent such accidents or incidents from happening again. All of this is the way to make use of "unexpected learning experiences".

The consequences of other people's mistakes help us learn. Brion K. Hanks, one of the authors of this book, makes use of these data by distributing reports to the crews in the field. Over the years, Brion has been amazed, and delighted, that his crews use these reports to initiate safety conversations and learn from the incidents with the expressed intent that they do not repeat the blunders on their current project. It is difficult to fool yourself into believing "this couldn't happen to me" when you read about it happening to someone just like you doing similar work. There is nothing like a tragic real life story shared from the heart to make a lasting impression.

Through examining hundreds of these reports, the general consensus is that both management and the human factors play dual roles in the "why" of accidents. Human error is reported to be a primary root cause in virtually all of the cases.

Did you nod your head in agreement with this statement?

Here is another point-of-view. Yes, people need to learn from their mistakes, but if your data indicates that human error is a primary root cause of accidents then it is a major red flag to a process-improvement professional – your organization has a primitive understanding of how to standardize a process. A well-designed process strives to eliminate the possibility of human error. The real root cause pervasive across construction is pushing process design into the field. Doing this forces the worker to make safety, production, and process decisions under schedule pressures when it should have been engineered in advance.[51]

[51] This point-of-view is discussed briefly in Chapter 8 - a potential systemic failure to understand safety and quality as a process.

Casting so much of the burden for success onto the field person can generate fear. And then we blame them or fire them for making mistakes that anyone would make without a fully defined process. And it could be prevented. This is where management's responsibility kicks in. In the future, fear will be driven out of the organization through preventative decision-making, process understanding and science.

Learning the Lean Six Sigma tool set will help you become process-minded, more valuable, and drive unproductive fear out of your organization. Exposing real root causes such as fear and pain can be very powerful.

Blame amplifies fear. When we finger-point at mistakes without deep consideration for the real root cause, then people fear for their jobs and that has consequences as previously discussed.

The elimination of blame (as practiced in Behavior Based Safety) helps improve accurate reporting and drives unproductive fear out of the organization.

What is the nature of your company's business culture? Will "the troublemakers" get fired for pointing out problems? If so, you will have more accidents because workers will be afraid to speak up. If honest disclosure is encouraged, then you need to be sure the workers believe it. It takes consistent dedication to prove yourself and even one misstep can tumble the trust you have built.

We must be aware of the consequences of unproductive fears in the workplace and drive fear out of the organization.

Consequences of Worker Reputations

Your reputation works like a bank account. If you deliver on-time it is like making a deposit. When you have accidents it is like making a withdrawal. But your bank account is being watched. If you have too many withdrawals (safety incidents) the cost of employing you goes up and you get paid less interest (hired for fewer new projects).

If your work record continues to be unsafe, then you become overdrawn from your reputation bank account.

Poor safety is a form of mortgaging or borrowing from your reputation. The outcome is fewer opportunities for you and with higher interest on them (less pay, more conditions, more assurances, etc). You may not even be able to work on public works projects.

Allowing too many accidents could translate into being "blacklisted" by prime contractors and clients. Few want to take on a worker that will drag down their safety records.

However, if you are constantly delivering on-time and improving safety you are making deposits and building trust with your reputation bank account. Below are real world examples of safety 'withdrawals' and 'deposits' that could affect your company's reputation:

- Favorable contract terms on warrants and assurances;
- Employee walk-out due to safety problems and the cost of delays;
- Fines for safety violations, the cost of delays and medical expenses;
- Higher insurance rates for workman's compensation insurance;
- Investors expect more time and effort from you in due diligence;
- OSHA investigates you more often and fines you more heavily than companies with better safety records;
- Investigation by the FBI for collusion or the local newspaper for chronic safety violations;
- You are not invited to bid on some U.S. government projects that have contractual requirements that expect proven effective safety training;
- You are found and cited for negligent training, which leads to a civil lawsuit;
- Outside observers report pattern of disregard for worker safety to OSHA, the newspaper, and government officials because they do not get satisfaction from your company;
- Strong safety record gives a competitive bidding advantage through lower workman's compensation and other insurance rates that can be negotiated on the contract;
- Awards and honors that keep you top-of-mind with decision makers;
- Repeat customers and referrals based on previous excellent safety performance;
- Ability to assure project completion and mitigate project delays based on reputation.

In the short run, many companies believe safety can be compromised. But in the long run, failing to take safety seriously will affect reputation, quality of job opportunities, and their bottom line.

Consequences of Choice

When viewed with all the distractions of deadlines, pressures, and management expectations, decisions can get muddled. You can master the consequences of choice if you remember that your reputation is a bank account. Carefully consider how your choices affect results. What you put in is what you get out. Good management cannot tolerate poor safety practices and neither can you!

It is easy to allow production needs to take first priority because it does work sometimes - if you get lucky. In the short run, it is an easy solution - until something bad happens - until the root cause of an accident is attributed to you or someone you supervise.

It can start with a craft worker error, but responsibility and ultimate accountability will fall on management and they will not want to associate with someone who costs them money and who hurts their reputation. Choices have consequences.

One day an ironworker said to Brion, "After working for a company who shall remain nameless, I became concerned about my own safety because they were not serious about safety. There were so many accidents on the project. So I 'drug-up' and quit... I quit because I did not feel safe working there."

Too many companies do not walk the talk and continue to compromise their commitment to serious safety practices. You do not want to learn those bad habits or be associated with that kind of accident-prone work. That ironworker not only demonstrated safety integrity, but also held his new crew members to a high safety standard.

Let's look at the consequences of your choices when you use premier safety practices. People live to work another day. Families are stronger, friendships deepen, and communities are more robust when workers go home safe each day. The company's bottom line wins because there is significant cost to poor safety, even if too many companies do not yet recognize this fact. The company's reputation wins. There are many other wins, too.

Recognition Consequences

Verbal appreciation and public recognition of safe behaviors have a powerful effect on people.

When safety matters as a part of corporate culture, there needs to be a way to recognize employees for their safety efforts. How do the best companies reward their people for correcting unsafe acts or conditions? Simple recognition - appreciation - can have dramatic results. Positive reinforcement is a major weapon in the safety arsenal when the battle is being fought to change people's unsafe mindsets and habits. This is a relentless battle we must fight day-in and day-out. Have the will to do your part. As a frontline supervisor or craft worker your positive feedback has strong value to your team.

The best way to correct most safety mistakes is to praise publicly and scold privately. This goes a long way toward sending two symbiotic messages. On the one hand, we show appreciation for any safety effort and on the other hand, we hold every person accountable for any safety failures. This is an effective technique. There can be no compromise and there should never be anything less than full commitment to consistent safety reinforcement.

Respect and consideration promotes safety.

Also, seek honest feedback from your peers about your safety performance. Ask questions about what they encounter and perceive while working in the field. This shows you care about people, especially when you act upon their concerns. Positive action makes the difference between being perceived as providing lip service or delivering a real demonstration for implementing safety solutions.

The Consequences of Success

When we finally learn how to reduce risk to 100% zero harm, does that mean you will never be at risk again?

Suppose you found a way to achieve 100% safety and improve profitability by 5% and this is now a stable gain. You understand what you did to achieve these gains so you know what to do to continue to achieve this success. Would you be tempted to bring in the schedule a bit, defer some maintenance or take a bit more risk in some other aspect of your job in order to squeeze out a little more work, salary, or reputation?

Today most business people would do so. Most workers would do so. Most supervisors would also do it. Workers and businesses must both continue to grow or stagnate, right? Change is a fact of life. To grow, we must risk. Of course it is a natural tendency!

You must realize it is an ongoing competitive challenge to keep safety at 100% zero harm while also advancing productivity, salary and reputation. The important point in all this is that 100% safety is a moving target and it will always be a moving target. Safety is always under construction. That is the nature of change and change is the nature of life and growth.

NOTE: It is an important point to realize that there is not going to be a silver bullet to fix all of your safety problems forever. Even after you achieve 100% zero harm, something will change – new processes, new materials, and new competitors - and you will be back striving to get even better.

It is a compromise, therefore, to ever be complacent about safety. You can be proud of your achievements. You can take satisfaction that you and your crew were safe. You can even swagger with professional authority and be ferociously gratified that you have achieved the pinnacle of success as it is currently defined. However, you must understand that you cannot let down your guard or think you have done enough and believe what you have achieved will stand the test of time. Change happens. Competitors become stronger. Society and bosses expect more. There is always more to learn. You must always be vigilant when it comes to safety. The fact is - safety is always a work in progress that requires sustained attention and commitment. It is a journey of safety mastery.

Knowing this, let's ask some long-term questions that reflect what a serious-minded leader might ask. Pay attention to the answers you think you should say and what is your automatic response. In this gap lives a world of useful information.

What is the most useful skill necessary to address safety in changing conditions? What skills must we develop to reduce risk continuously AND improve the quality of our work AND produce more work?

What is the relationship between production and safety?

What is the relationship between greed and safety? What is the relationship between improved efficiency and safety? How does competition with our peers influence worker safety? What is the relationship between job performance and safety performance? How do layoffs affect safety?

Do we have a hidden mindset that affects safety; are other crews less important than us?

Is there a code of belief in the work environment that leads to neglecting worker safety? Can you think of a situation that would justify a supervisor compromising safety? If so, what is the situation that might warrant it?

How will adopting new technology affect safety? How about using new equipment or adding a new subcontractor to the job?

Safety leaders must have the ability to recognize critical issues. We all must be flexible enough to change.

Summary

If by now you have accepted responsibility for the safety of yourself and your team, you have joined an elite group of leaders.

Even if you achieve perfect safety at one moment in time, things will change and you must be flexible enough to look constantly for new ways to be safe. Using today's mindset it is so easy to focus on other things before safety. It's simple to think you are already doing enough and it feels natural to put safety on the chopping block to be compromised when things get tough.

That is the current situation.

This will change as leaders and workers begin to understand the fundamental contradiction of their industry, quick-vs.-safe, and the extraordinary opportunity it presents to improve safety and profits. But it means that workers and supervisors must be more responsible now. You must modify your time-honored self-image of hard working (physical labor), do as you are told, get 'er done men and women. You must become just as hard working and hard thinking, get 'er done safely men and women. You are going to have to include learning and rigorous problem-solving in every aspect of your job.

Realize that improving safety will be the key to making that happen. Safety in the workplace cannot take a backseat to anything. The priority must be clear and present - constantly out-front.

The kill rate in construction is horrendous (this is a war term, but it is appropriate). To do nothing is to get more of the same and that ought to be totally unacceptable to any sane person.

Individually, and morally, we hope that you are tired of the negative consequences of prioritizing people below other values, which sponsors too many senseless accidents through the unconscious avoidance of pain. Accidents can be stopped. Through the power of your will, integrity, unwavering commitment, and problem-solving prowess, you can make dramatic changes that will improve safety success.

If that rationale does not serve you then stop being a fool. You can make more money when you accept responsibility. Poor safety costs you money and opportunity. If for no other reason, choose to be responsible because it will help you keep your job and help your company achieve greater profits.

Bottom line #10: ***Accept heartfelt responsibility for safety because positive safety consequences come from positive choices.***

Safety Mantra #10: ***Choose well. Actions have consequences.***

Take-Away Points: ***Chapter 10***

- You are responsible for the consequences you create. Every action or inaction produces a cascade effect.
- Be more preventative. If you want to control safety you must eventually control the process. If you want to control the process you must understand it.

- Today we expect crafts people to do the job of a process engineer, under the pressure of schedule and without time in advance to think.
- Learn new skills so you have different methods to solve issues. Science, Six Sigma, problem-solving skills and other approaches can resolve safety issues.
- Making mistakes is not good, but learning from them is great.
- Safety should be built into every process.
- Blame amplifies fear. Drive unproductive fear out of the organization.
- Your reputation is like a bank account. Delivering quality work on-time makes deposits. Accidents make withdrawals. Make large deposits and no withdrawals to grow your reputation.
- Praise publicly and scold privately.
- Respect and consideration promotes safety.
- Safety perfection is a moving target and it will always be a moving target.

"Tell me and I forget. Teach me and I remember. Involve me and I learn."

- Benjamin Franklin

"We learn by example and by direct experience because there are real limits to the adequacy of verbal instruction."

- Malcolm Gladwell, Blink: The Power of Thinking without Thinking

Chapter 11 – Best Practices Virtual Walk-Through

This chapter is comprised of a virtual walk-through in a day in the life of Brion, one of the authors of this book.

Brion drives onto the site, passing a sign stating Kiewit's First Rule of Safety, "Never walk past an unsafe act or condition." Brion reminds himself as he heads to his office, safety is an important and pervasive mindset. As safety manager on this project, it is vital that he set the tone of the worksite for every employee and subcontractor at every level, every day.

"I may be the Project Safety Manager," Brion states softly under his breath, "but everyone on this site is empowered to at least call attention to an unsafe condition." This is one of Brion's Safety Mantras that is shared daily. He never forgets it.

Brion is onsite early, as usual. He unlocks the door and in his mind's eye he scans how he left the site yesterday. He's thinking about those subtle, often overlooked issues: was there any uncapped rebar, leading edge issues, hand/guardrails missing, unprotected open hatches, access issues, or VIPs onsite without proper PPE in place? No; in large part because workers know that they also have the responsibility and are empowered to do their part.

He looks over the day's work schedule for any meetings or training requirements, makes notes, checks emails, and gets things ready for the day. He was asked to be safety manager on this project after it was already under way, so he had to get up to speed quickly. His preference is to be involved in a project's Readiness Review and development of the project-specific Safety Plan. "It's better if I'm there at the onset of the project to set safety in place as a foundational aspect of the job in ways that I have found effective and which sets a safety tone requiring rock solid commitment," states Brion to this book's co-authors, and he goes on to say, "I like to make sure that key people in key positions know to keep safety first. When this is accomplished, the attitude of safety ripples outwards in a very positive way..." And so another day begins.

Brion greets the supervisors and foremen as they arrive before their crews and he observes their pre-work meeting. Today he conveys a safety tip that he wants shared with all crews.

Soon after craft people arrive, they form into crews and meet for the morning Stretch and Flex, with all their PPE on, which is followed by work plan assignments. Brion knows which group he will Stretch and Flex with today. He understands that he needs to be visible to this particular group. He'll continue to build relationships and impart important safety information. They spend 10 to 15 minutes in group exercise, warming up the body, waking up the brain, getting the kinks out, and bonding. People always tell Brion they are skeptical of this practice of morning exercise when they first hear about it, but after a few days, they admit it becomes a vital part of the day. Stretch and Flex promotes clarity of thought, improves flexibility, and group cohesion. The benefits of this practice are statistically significant. Depending on how the day goes, they might do an abbreviated version of Stretch and Flex after lunch as well.

After the Stretch and Flex, the supervisors and foremen go over this shift's work. If necessary, Brion raises his hand and shares new information or discusses something he wants to bring to everyone's attention in order to get it fresh in their minds. He reviews any incident alerts or "Fatalgrams[52]" in the form of a proactive lesson. Brion's intent is that this lesson-learned approach will prevent workers on this site from experiencing a similar tragic accident.

Now it's time for every group to discuss their relevant "Play of the Day". They spend as much time as necessary, which sets each crew up for daily success before any work begins.

- Once a month there is also a mass meeting for all project personnel, including subcontractor employees that Brion (as Safety Manager) leads.
- Every day, crews fill out a Daily Snap Shot because "Safety is no accident." Also daily:
    - Foremen are required to "Ask people in your crew questions that are relevant to your operation today (biggest risk, toughest task, little 'stuff' that could hurt us, basics of their operation).
    - As a part of engaging their crews, the foremen also ask, "How can we improve our access? How can we improve our housekeeping? What tools and equipment are we using today that could hurt us?"

[52] Fatalgrams are alerts from the Mining Safety and Health Administration about the tragedy of fatality accidents and their root causes.

- There are reminders such as: is your Job Hazard Analysis current and is the JHA reviewed and signed by each person on the crew?
- Daily visual inspections are completed for all equipment in use.

- Once a week there is some type of tool training.

After the crew begins work on the project, the safety manager walks the site for one to two hours in order to connect with people and look for safety issues (is someone using a cutting torch without a fire extinguisher nearby; is that an exposed extension cord across a walkway, and is there water available to avoid dehydration especially on cloudy days when it is often overlooked, etc.). This is a chance to have Brion's presence noticed, build relationships, and observe the reality of work operations. If he observes a problem, he makes corrections on the spot with an eye towards Behavior Based Safety. During the walk around the site, the safety manager takes photos for other meetings throughout the week and to place in the safety folder for the project.

Of course, Brion is always available via the radio for any issue that requires his immediate attention and expertise. On this particular site, there are four tower cranes and four lattice boom cranes. They use the concept of vertical height separation as well as anti-collision software as best practices for safety. There are also dedicated channels between operators and bellman. During a lull in action, Brion reminds one bellman that he and his spotter must maintain visual contact (to signal) and/or proximity for verbal communication. Constant reinforcement of this rule is important. The results of an incident involving cranes could be catastrophic.

Generally on Monday and Tuesday the safety manager conducts a new-hire safety orientation and performs the drug screening that is done for every new employee, both craft and staff. Providing a positive first impression for new-hires is invaluable to set the initial safety tone for the project. During the orientation, PPE is handed out. Brion presses one point repeatedly, "We are serious about safety. We care about you and want you to go home safe every day. We will hold ourselves accountable. And so will you."

Brion also stresses three additional points, "All workers will be given the correct tools for every job. And you are expected to use them. If you do not have the right tool, take the time to get it." The second point is, "All people will abide by company policies. Do not take shortcuts even if there is time-pressure to get a job finished."

The third point Brion communicates to the new-hires is their value as a fresh set of eyes. Because they are not yet caught up in "this is the way it's done here" mentality, they can give invaluable feedback. Brion tells all new-hires, "We want you to look around where you are going to work; we want you to ask questions about your work conditions; we want you to speak up. The JHA is a 'Living Document' and needs to be updated should you see some hazard or changed condition that could adversely affect crew safety."

This particular project requires anyone new to the site to wear a red hard hat for thirty days. This immediately announces to everyone that you are new and may need to be watched to make sure you understand the project's safety culture. This allows all employees to keep an eye out for the new hires. As well, Red Hats are not allowed to work by themselves. No one is above these rules. When Brion joined the project, he also wore a Red Hat and he was the safety manager. That set a good example. When one of the workers mutters that he does not like wearing the red hat, Brion jumps on the opportunity to adjust the perception. "There is no shame in wearing a red hard hat. It simply means that you are new to the project and it makes it easier for all to observe and watch out for you. But I understand what you mean. I was happy to graduate to a white hat."

Once a week, the safety manager leads a foremen's Safety Meeting. Lunch is provided, which keeps people focused on the subject at hand, allows for no excuses, and builds team working relationships.

Brion performs a myriad of other jobs during the day including: analyzing the projected tool needs and inventories safety supplies. He orders and distributes safety supplies as necessary and tracks where those safety supplies are used. Then he logs the use to the appropriate cost codes for given operations. Brion never misses an opportunity to reinforce the "right tool for the job" policy.

As Brion performs daily safety-walks he is always looking at access to all operations; schedules anything and every thing to do with proactive safety, and watches the project from all angles, especially where lines of work intersect.

The safety manager is often called out into the field during each day to view issues, provide a clarifying point-of-view or confirm a company policy and sometimes to investigate an incident. One of Brion's famous quotes is, "When in doubt, don't!" He continually reassures workers that it is okay to speak up and ask question when they get a "gut feeling" that something is unsafe. If tensions are high, they call Brion. This puts the responsibility of the safety vs. production contradiction firmly on Brion's shoulders and allows the crew to work without fear of reprisal. Brion stresses that if there is ever a doubt about a safety issue, immediately contact him as safety manager to resolve the issue. "Ask questions," Brion authorizes; "Let me make the decisions. And if it affects the operation, then I will include your supervisor or elevate it to the Project Manager as needed."

When a project is under time pressure, crews will often work long hours or there will be multiple shifts. When there is a swing shift, the starting shift ends at 5:30 pm and the Night Shift begins at 4:30 pm. This allows Shift Two to have their Stretch and Flex and attend their Play of the Day meeting with the Night Shift Supervisors and foremen. New Hires for the second shift receive their orientation, too. There are no shortcuts; nothing is minimized or left out.

A couple of times a week, the safety manager performs training such as fall prevention training, designator operator, rigging and signal person training or confined space training.

A major part of Brion's job is to perform first aid when needed. He has already established which medical clinic will be used and he escorts any employee there. He also takes care of all required paperwork and documentation related to safety and performs pre-employment and post-incident drug screening. Brion believes this is vital to safety success on any project.

Brion tells all workers about a voluntary policy for medical information such as allergies or physical conditions (for example, allergy to aspirin or bee stings, or having epilepsy or diabetes). The personal information is placed in a sealed packet on the inside of the person's hard hat. If any of the workers should fall unconscious, Brion or another supervisor can quickly determine first medical steps or eliminate options that might hurt the situation. As well, he can contact the appropriate doctor or relative if the employee indicated. This can save lives while maintaining worker privacy.

Brion always knows what the weather forecast is for all shifts. He even keeps a lightning detector in his office for monitoring any incoming storms. When the cranes are being used, he monitors the wind. This way, everyone follows the manufacturer's guidelines for crane use.

Safety has momentum. So does chaos. Brion makes sure that the people in charge have in-field experience because each project is only as good as the speed of the leaders. When key positions are populated by key people, other workers will follow their cues. The only way to stop the momentum of chaos and accidents is to have a dramatic pattern interrupt and then begin again with the key examples and leadership in place.

If there is one thing Brion has learned during his 40-plus years of construction experience, it's "be consistent" because no one is above the law. When Brion holds every person accountable in the same way, it's good for morale. As Brion walks around the site he gets the pulse of the work environment. When he witnesses solid safety success, he's filled with pride that crews are working safely and lets them know. His goal is to build a safety team mentality always where all workers contribute daily to safety and where each supervisor is fully aware of their team's capabilities and limitations and is held accountable for the team's actions.

Today, there have been no incidents and Brion continues to remind everyone that there is a strong project safety culture in place due to each person accepting responsibility for their own safety and the safety of their co-workers.

Bottom line #11: ***Live safety best practices. Focus on them. Dedicate yourself. Embody excellent safety actions and choices in each moment.***

Safety Mantra #11: ***Learn from the masters to create your safety foundation.***

Take-Away Points: ***Chapter 11***

A day in Brion's life as a Project Safety Manager requires that he be a living example for all employees. Here were some of his daily duties:

- Perform employee drug screening and safety orientations √
- Help with Job Hazard Analysis reviews in the field √
- New operation start-up safety meeting √
- Project-required training √
- Safety inspection tours √
- Mentoring, coaching and leading √
- Goal setting and accountability √
- Praise people publicly and scold privately if necessary √
- Manage project-specific safety programs √
- Subcontractor safety management √
- Communicate with and respond to any Client concerns √
- Manage and document any incident √
- Ordering and taking inventory of safety supplies √
- Manage the Safety Recognition Program and update the log √
- Show you care but be firm when necessary in order to keep everyone in line with company policies√
- Ensure management involvement at all levels regarding safety √

Workers, safety leaders, management, and frontline supervisors must walk-the-talk of company safety priorities. This must be done so well that every person in the workplace embraces the commitment to "nobody gets hurt". And while tomorrow will be a new day, the goal remains the same: everyone goes home safe every day. That is safety perfection!

PART III – PEOPLE DO WEIRD THINGS

> *"Nothing in the world is more dangerous than sincere ignorance and conscientious stupidity."*
>
> *- Martin Luther King, Jr.*

> *"Only one who devotes himself to a cause with his whole strength and soul can be a true master. For this reason, mastery demands all of a person."*
>
> *- Albert Einstein*

No excuses. You have a critical and important role in reducing the number of deaths and injuries on construction worksites. And so we begin Part III with this quote,

> *"Safety is a part of every person's life, both at work and wherever else we may find ourselves. In fact, safety permeates everything we do. Because of this we cannot, and should not, ever allow safety to take a back seat in the course of our daily lives."* [53]

"Yeah, right…"

Are you thinking it? Or did you gloss right over that quote because the content was expected? Admit it if you did because the information about how you responded to the quote is important. Aren't there times when each of us use our attention in cruise control or allow unfocused cynicism to prevail on the job? These are both serious challenges that affect safety and are two issues that need to be addressed. Make note of your reaction because it is critical information that you can refer to as we read Part III.

---

[53] This visionary quote was created by Brion K. Hanks, one of the authors of this book.

If you ever doubted the value of pushing for continuous safety improvement in the workplace, Part III of this book will explain in detail why you are conflicted. Although you may recognize some of the information in this section, our approach explains it in an original way so you understand why sometimes safety is your best friend and at other times you act as if it is your worst enemy on the job.

So look inside yourself and watch your reaction as you ask this question and then think:

Can you achieve an accident-free record on the job? *[Please, take your time to think.]*

Now make a few notes of your reactions. Did you react with a single feeling or single answer? Did it require multiple answers? Was it positive or negative? Did a dialogue begin inside? Did your mind go blank? Did you size it up as a challenge or see it as a big problem? Were you instantly cynical? Did you feel motivated, optimistic, burdened, or pessimistic? Did you feel, "Yeah, that's possible" but then no other arguments or internal dialogue appeared? Did you notice that you do not care? For now this is just information with no right or wrong answers.

People's reactions to the possibility of an accident-free construction site is often complex. At this point you may categorize yourself as ready-to-help, skeptical, motivated, unmotivated, cynical, optimistic, or any number of other labels. We put a lot faith in our feelings, assessments, and choices and that is both a good and bad thing to do.

Our faith is both good and bad for the same reasons: We know who we are, we believe in how we feel, and we act in accordance with our beliefs - well, most of the time. As you will see in this part of the book, people do some very strange things:

- We make excuses when we know better.
- We say things we don't mean.
- We follow a leader that we know is harmful to safety.
- We don't pay attention when it matters the most.
- We bow to group pressure when we need to take the lead.
- We compromise when we know it is foolish.
- We risk our valuable lives for a few minutes of speed at work.

If that is not strange enough, it gets even more interesting: when we are offered a genuine and sincere approach to improving safety; some of us become suddenly confident that nothing more can be done and mock positive efforts to improve. Go figure.

If you have not noticed, there is a process that we just described. The process of facing a question, glossing over feelings, making rapid conclusions without a sincere examination of evidence is called an automatic response. And we have a lot of them that interfere with safety.

People do weird things when their automatic response is not a good match to the situation. Add to that stress at home, a new job, brain style differences, medical conditions or a demanding boss and guess what? People will do weird things.

Our goal in Part III is to expose the hidden triggers, the sequences of events and automatic responses that bring harm to you on the job. We will also discuss how to handle these issues so new safer patterns in behavior and response can evolve. We are not going to pamper you. You are going to have to think, feel, and stretch. As far as automatic responses are concerned you are the student, the red-hat, the new-timer, and probie on the job and you need to earn your way.

You also need to reconsider the notion that accidents are *unexpected* and *unintentional.* Here is why. Once you learn that people do weird things because of hidden triggers and automatic responses and we together identify these triggers / responses, then it can no longer be called unexpected. We need to regain control of our automatic responses so that our hidden intentions and expectations do not express as dangerous choices.

"The past does not have to equal the future." Tony Robbins

We reaffirm our dedication to saving lives now by understanding the weird things people do on the job and learning what we can do to create safer more appropriate responses.

"The human brain is a complex organ with the wonderful power of enabling man to find reasons for continuing to believe whatever it is that he wants to believe."

- Voltaire

"I was texting my friend a warning, 'Be careful, the road is slick.' That's when I slid into a ditch."

- The excuse given by one of the author's nephews

Chapter 12 – Ignorant Things We Say and Believe

It is time to expand the brief introduction about ignorance and assumptions provided in the basic concepts section and to generate some lessons from real life scenarios.

As each of us heads-off to work today, we have every intention of coming back home alive after work. We have been trained and know how to behave to ensure our survival. But there are a few things we often think that may undermine our safe intentions and cause serious harm to ourselves and others. They are:

- Nobody's looking.
- I have to; I have no choice.
- Accidents happen.
- Rules are for fools.
- It's a calculated risk.
- It looks safe to me.
- It's good enough.
- That won't happen.
- It's not the right tool but it'll do.
- It will only take a minute.
- I've done it this way for years.
- I knew better, but …
- It's just a short distance.
- I know more than my foreman.
- I don't want to cause trouble.
- Nobody told me.
- It is easier to do it that way.
- My boss told me to do it.

It Will Only Take a Minute / My Boss Told Me to Do It

EXAMPLE ONE: Here is a real-life story about "It will only take a minute" and "My boss told me to do it". Approximately 25 years ago, on a sewer project there was a trench that was 18 feet deep. A tool had fallen into the trench and a superintendent told a laborer to go down into the trench to retrieve the item. At that moment, this was an unshored trench and the laborer did not want to do it. Intuitively, he knew it was dangerous.

In front of his peers, the superintendent said, "Go on, do it, it will only take a minute." Basically, the superintendent shamed the laborer into going down into that unshored trench. As the superintendent walked off he told the other workers, "Tie a rope around him; that way if the trench caves in we will be able to find him."

How about that as a real-life, blatantly callous disregard for the safety of an employee directly under his charge? In our minds, this is nothing short of a criminal act. Do you think it is okay? At a minimum, it should require immediate termination of the superintendent. This sends a clear message and sets a strong demonstration that such behavior is not tolerated. Any less punishment, in our view, makes the employer complicit in the "criminal act" as it appears to condone the superintendent's actions.

This kind of all-in attitude is what is required for a zero harm safety record.

The worker went into that unshored trench even though he knew it was dangerous because, "He didn't want to cause trouble" and his "boss told him to do it." This is another dangerous compromise based on fear and pain.

Many of us have seen scenarios where there is a blatant disregard, disgust and cruelty displayed toward laborers from supervision and management - those who are supposed to have our back. As we analyze the lessons from this event, we realize that supervisory hostility and complicit co-worker behavior makes it very difficult to handle this issue.

Do you believe the construction companies you work for want managers like that on their jobsite? Could you stand up against this kind of pressure and blatant disregard for human safety? If you saw this, would you correct it or put your head down and pass by this unsafe act and condition? We started with this example to show that we understand how difficult it can be to make safety a priority. For a moment take a breath and imagine what this worker might be thinking in this situation:

- This is a risk, but not every unshored trench collapses. I might be okay.
- Since 1984 there have been about 1,941 injuries and deaths related to unshored trenches as reported by OSHA. I could die.
- Everyone is afraid of the boss; he is reckless. My safety is not that important to him. If I get hurt, who will take care of my family?
- I've seen what happens to people who make trouble. Management is hostile when workers speak up. What will my wife say if I get fired?
- If I report this, I'll have to go all the way. Who would I talk to? I'll have to have my facts straight. What if they don't believe me? Would any of my co-workers back me up? I'll ask a co-worker to shoot a video with his phone while I argue with the boss.
- I'm scared.

To be successful you will need to find out a few things. Start by asking the first question you can think of. Pay attention. As you answer each question, the next possibility seems easier to address and not as intimidating. As you continue to ask and answer questions, the path that you would like to take becomes clearer.

- Does OSHA have whistleblower protections? How do people whistle-blow? I'll call OSHA or check the web.
- Does my company have a safety ombudsman? I can have a friend look into it for me.
- Who in upper management would I speak with? Would management even listen to me? Who is the safety manager or the highest safety person in this company? What is her name? What's her contact information?
- How would I present this to my upper management? Can I formulate a strong plan that management will respect? I will need to work with others to make it happen.

Notice how fear can hold you back. Also notice how facing each fear adds clarity and creates a path to the solution. This is your path of responsibility.

Also notice how intimidating it would be to try to address all of these issues in the dangerous moment. Instead, prepare in advance. Ask questions. Acquire the stated safety policy. Discover who to contact if someone asks you to do something dangerous. In a hostile environment, talk with other workers on the job - who would be likely to back you up and who you would back up. Create a plan before one is needed. If you never have to use it, no harm is done and it will make you feel more competent should a situation like this one arise.

For now we will stop here, but realize it takes full commitment to make you and your peers safe.

LESSON: You must be all-in for your own safety even when your management is not. To do this you must identify the problems you face and solve them one-by-one.

Let's move on to the next example.

I've Done It This Way for Years

EXAMPLE TWO: Driving fast on a curvy road (dangerous thinking and the habit force - "I've done it this way for years.")

This example is based on the thought patterns, "I've done it this way for years" and "That won't happen to me". Driving fast is very common and curvy roads can be fun - so relax your resistance and admit you have done this or at least wanted to do it. It is easier to face the truth without resistance when everyone has done the same thing. Yet, these patterns of ignorance and assumptions are dangerous thinking.

Here are some common excuses for driving on a mountain road too fast: "I've driven this road a thousand times; I know the road like the back of my hand; I could drive this road with my eyes closed; I've never had a problem driving it before; and everyone else drives it fast, too."

Here's how the pattern of thinking unfolds. It may lead to disaster or you may be lucky for years:

- I've never been hurt doing this before (after a few times of thinking this, the thought turns into)...
- It's safe to do this (which, in turn morphs into)...
- I'm comfortable doing it (and this complacency turns into)...
- Since I am so competent, I can go faster and push the limits a bit more. I'll just use a bit of the shoulder, cut a corner or two, and use just some of the other lane as I drive.

This is the momentum of "false learning" in action. The momentum of habits and incorrect thinking does not change the statistical potential for an accident. It remains a risky behavior no matter how many times you have successfully driven the road. Let's look at the possibilities from another point-of-view:

- Time-based: You usually drive fast on this road at night and you've never seen anyone walking on the side of the road so you clip the shoulder line on a regular basis. Today you are making a day trip. Because you have never before experienced a person walking on the shoulder, you are horrified when you round a corner and see a man walking to his mailbox. You have no time to swerve.
- Space-based: Just because you are "mostly" on your side of the road, does not mean another person with your same sloppy mindset won't also be driving fast and using up a bit more of the road than is their share. You both round the curve and find you are occupying the same place at the same time at high speed. It happens. Brion K. Hanks, one of the authors of this book, is personally aware of a double fatality accident when two semi-trucks rounded a corner and hit head on.
- Condition-based: Deer are usually in the forest; today one is crossing the road. It is the first heavy rain of the year and the road is slick with oil; or someone's tire is flat and there is no place to pull off the road so they are limping down the hill at 5 MPH just after a blind corner. A tree branch fell off the tree and lays across the road; or a rock slide covers your part of the road; or earlier today, a small earthquake hit this area and half of the road is

missing. You have never met these conditions before so you are going too fast to stop in time.

Now let's go back to construction. How are these thinking patterns consistent with your jobsites?

LESSON: It is ignorant and reckless to deduce your skill has anything to do with successfully facing a risk when you are ignoring safety solutions. Yet it is one of the common, dangerous and weird things people believe - that their experience in "beating the odds" reflects their skill or understanding. When, instead, it is the statistical luck behind the risk.

If the risk is 1 in 1000 then you may experience 100% successful outcomes for a long time. Most people don't realize that their skill is factored into the risk already.

EXAMPLE THREE: This example is similar to the previous example but the odds of harm may be much lower. Notice how your reaction may be different.

Brion observed a carpenter changing out a rotary blade on a power saw. The power cord was still plugged-in. He quickly interceded, "What are you doing?"

The carpenter replied, "I'm putting a new blade in my Skil saw."

"But it's still plugged-in," Brion pointed out.

The carpenter said, "Well, this is the way I've done it for years, and nothing has ever happened."

"Hmmm," Brion murmured and then he asked, "Do you know that you are violating company policy?"

"Yes," the carpenter replied.

This form of ignorance is the same as previously noted. The carpenter thinks that his previous successful outcomes of changing the blades indicate that he is not at risk.

But the lesson that we will emphasize here is the negative influence on other team members as he is observed doing this without correction from his peers. He is showing a public and willful disregard of a known safety risk. Impressionable co-workers will likely pick up his habit.

As a part of accountability, this carpenter was given one day off work without pay because he knew better and did it anyway. Why does it sometimes take being reprimanded and penalized to make us change behavior when we know better?

If the consequences are more severe if the carpenter offends again, then he will probably start being careful to comply. We should be grateful if our supervisors or safety managers help us overcome our dangerous beliefs in this way. This kind of Behavior Based Safety interaction does create a safer environment.

LESSON: Disregarding safety policy can hurt the team and make you all unsafe. Don't wait to be corrected to prioritize your safety and that of your peers. Take those few extra moments to work the safe and correct way.

EXAMPLE FOUR: Here's another real-life account of a worker violating stated company policy and an OSHA regulation (committing an unsafe act).

The subcontractor employee was higher than six feet above the ground with no fall protection or hand rail and walking along the leading edge of a newly constructed wall. Brion asked the man, "Do you realize what you are doing?"

The subcontractor's employee gave Brion "the deer in the headlight look" and swiveled his head trying to figure out what he was doing wrong - he had no idea what the issue was. Brion pointed out, "Do you realize that you're walking on top of a wall that is 8 feet off the ground?"

The worker ruefully said, "Now I do."

Herein lays a challenge for companies that use subcontractors who don't train their employees as well as your company trains their employees. Even if these workers learn safety on your jobsite, once they leave and work on other sites where safety accountability is not in place, and perform their job in unsafe ways and are never held accountable, they may forget or go back to their old unsafe ways. If they are rewarded for ignoring safety with the belief it will deliver work faster, they may develop unsafe thinking patterns.

Often the response we have on the job is to say, "Nobody told me." And while that may be true, when it comes to safety there really are no excuses that justify it. A good response at this point would be for the worker to ask some clarifying questions and take the lessons to heart.

Of course, part of the answer is for the company to establish policies clearly, rigorous safety requirements and then train, train, and re-train their people prior to any work being performed.

But the other part of the answer is that it is your profession and your responsibility to get the training you need to work safely. In fact, you can take the notion that you will be better trained and more knowledgeable than your peers. Think about how valuable you can be on the job if you know the OSHA and company safety handbook as well as any inspector.

The big challenge was stated in the previous example: when unsafe acts are repeated over time without any consequences or accountability, the wrong behavior is self-reinforced, is reinforced for your peers who see you, and for those you train. Your unsafe acts live on even after you are gone. Unsafe actions gain momentum and influence future behavior in a destructive way. You don't get a "do-over" if an accident happens.

We humans have a tendency to believe accidents will not happen to us. And let's face it, some of us have hidden motivations like the rush of beating the odds - similar to the rush gamblers feel. The benefit for shortcutting a safety procedure might earn you praise at getting the job done quickly - might earn you a reward every time you are quicker than others who are doing it the correct way - until you pay the consequence. It might not happen for years or may never happen to you. But it will happen to someone and you may never realize that you contributed to the root cause of someone's death.

LESSON: Our excuses and our unsafe actions are transferred like a virus to our peers and live to infect future employees. Be resolved that you will be a professional and knowledgeable example of safety.

I Knew Better, But …

EXAMPLE FIVE: An electrician was going to work on a building's electrical wiring. He felt he was being safe as he took his voltmeter out of his belt and checked the wiring with his meter. It had no positive reading so he reasoned the power was off, as it was supposed to be. He climbed four feet up a ladder and grabbed onto the electrical wire, which turned out to be live.

The electrician immediately knew he was in trouble because he could not let go of the line. Being very experienced, he thought he might have one chance to live. He pushed himself off the ladder and gravity literally pulled him free of the live electrical line.

He performed a root cause investigation of the accident and discovered that the batteries in his voltmeter were dead. That's why there was no positive reading when he tested the live electrical line. This is how the electrician related the story. And the story seemed believable to those that listen to his story because even after twelve years, when he speaks of the incident, he is visibly shaken. This man knows how close he came to dying.

Prudent safety procedures were put in place based on the electrician's story, which is test the voltmeter on a known live line before checking a line that is supposed to be dead prior to commencing work. However, after analyzing the facts, the electrician's story shows some holes. Let's examine the evidence:

- Analog voltmeters can read voltage without a battery. Voltmeters do not rely on a battery to operate.

- Digital voltmeters need a battery to power the display to read voltage. If the battery was dead, the electrician should have notice the display was off. It is very obvious.
- All wires should be checked before touching them; power boxes should be shut off and locked and tagged out.

After looking at this evidence, we strongly suspect this story is a fabrication and that the electrician did not check the lines. In many root cause investigations a false mea culpa is claimed to avoid severe punishment because what they did was plain stupid.

The consequence of this lie is to pass on false safety knowledge, which could be dangerous to the company culture. It could be the electrician did think the story was true and was basically incompetent. Brion's[54] Dad often says, "Do you think or just think you do?" There might be another explanation and the electrician just thought this sounded reasonable. Maybe there was a language barrier.

Just because a worker is experienced, does not mean you must take everything they say as absolute truth. There are lots of reasons why people say things. It is critical to include rigorous logical thought and examination of issues when it comes to safety.

LESSON: Trusting stories over science translates into less productive and unsafe work patterns. The standard of proof for safety practices must include science and clear thinking. We must be rigorous and professional in our use of knowledge.

Excuses are Patterns of Ignorance and Assumptions

Despite all training, humans make false assumptions, rationalize on occasion, act wrongly, and are subject to overconfidence. Errors happen and then we are very sorry for our choices, but by then it is too late.

Are there other patterns of ignorance and assumptions that we should be aware of - patterns that all of us humans will indulge in on occasion - patterns that may undermine our intention to go home safe everyday? These patterns often express as excuses for not being observant, for not thinking clearly, for avoiding responsibility, or for not knowing. Notice this as you read them.

"Nobody told me…" is an excuse in which many people indulge. Perform safety orientations for all newly hired employees and clearly spell out all of the safety do's

[54] Brion K. Hanks, one of the authors of this book.

and don'ts. Ask questions during these orientations. Make sure you understand everything you are taught.

Another excuse often heard on construction sites is "It wasn't the right tool but it was handy so I used it anyway." Part of safety success is conveying to workers that they should always get the right tool for the job. And reinforce that the company wants workers to take that little bit of extra time to get and use the correct tool. Workers who are doing a good job are workers safely using the correct tools and equipment.

Ignorance by Word and Deed

EXAMPLE SEVEN: "Accidents happen" and other flippant responses are typically made by someone who does not take responsibility and shows that safety is not taken seriously. This particular excuse is ignorance in action and is totally unacceptable. This type of person holds their mind in a state of deep unawareness and displays an ignorant resistance to learn.

- Just because you dodged a bullet once while performing an unsafe act, does not mean you will dodge the bullet in the future.
- Just because you dodged a bullet does not mean your example didn't cause someone else to try the same and get hurt or killed.
- Just because you didn't get hurt does not mean your sloppy actions didn't cause someone else's accident or cause rework or ...

The next generation safety program will not tolerate this type of faulty thinking.

Today, unforeseen incidents do happen; but once we are aware that they could happen then we are responsible for making sure they never happen again. Because we understand that accidents are preventable through the application of knowledge and right action, it is totally unacceptable to hear someone say, "Yeah, that happens sometimes; too bad." This type of person holds their mind in a state of deep unawareness and actively denies that accidents are avoidable, thus they display an ignorant resistance to learn.

EXAMPLE EIGHT: A rebellious worker once said, "Rules are for fools." This is such a ridiculous statement, it hardly warrants discussion. However, it is interesting to ask, "Why do people rebel?" Is management giving mixed messages? Is this worker simply contrary? If so, is that appropriate for the job?

Some people don't like authority in general; some workers don't like a particular person in authority; some people don't like to be told what to do; and some people just flat don't care. The person who says, "Rules are for fools" does not see the value in the rules being imposed upon them. That type of person is dangerous both to themselves and to co-workers.

Excuse-based thinking patterns need to be changed quickly (for short-term safety gains) and mined for their precise meaning so problems can be solved in the long run. Without engagement and sincerity, the individual displaying a "rules are for fools" attitude must be removed from the workplace. It's not complicated.

Here is the last excuse that we will share: "I took a calculated risk…" Knowing better and then doing the risky action anyway puts people in danger. Doing it over and over again is similar to playing Russian roulette. Do it often enough and it will get you!

At-risk behavior is not only foolish, it's unacceptable. At-risk behavior must never be tolerated. To this end, clear expectations must be set. Accountability needs to be conveyed and demanded with consistency so that workers get the message and live the message. It is every person's responsibility.

Ignorance and assumptions often show up as a belief in past experience. For example, it's like driving fast on a mountain road - drivers may think it is safe because they have been doing it for years and never had a problem. A woodworker might think it is safe to "change my saw blade without unplugging the saw because my father did it that way for years and he never had a problem." If you look closely at the logic behind an experienced worker who routinely "dodges bullets", you can see that it is really not logical. Construction workers need to step up to a new level of expectation for themselves and get out of the mindset of making excuses. They must start taking responsibility for safety.

But, let us just introduce the idea that this mindset is only half of what is needed. Inventive thinking and resolving contradictions is a major gap that we also see in construction safety.

As we summarize this chapter, we need to check-in with you. It can be very difficult, especially for a construction person who is used to rapid action, to think beyond what is necessary - a good solution. If you really don't want people injured or killed on your jobsite then find the way to get over this way of thinking. You have to accept that you must become more and we will be putting you through the wringer to get you there. If you engage with this book sincerely you will experience discomfort and will grow. Use the pioneering courage that is your heritage. We are grooming you to become a more powerful construction professional.

In the current situation, let's bury, "I took a calculated risk…" in all its various forms. It is probably the most insidious and persistent thinking pattern that is a roadblock to total accident prevention. Having said that, it is important to clearly understand the underlying benefit of that thought process if we are going to successfully overcome it. Is there more to learn? Yes. Clear and inventive thinking will show us the way. You see, the fact is these thinking patterns persist despite attempts at rooting them out.

There is a reason. Now, just because there is a reason that has lasted centuries, does not mean it is a decent reason. It just means that no one has yet understood enough to root it out. We are going to learn more during Part III so you can begin to root it out of yourself.

It will take genuine effort to achieve safety perfection. Right action is required by everyone who desires it so badly that they will do their part 100% of the time. Make it happen!

Bottom line #12: ***A hidden mindset of risk and recklessness leaves a trail. Look for evidence in the form of ignorance, assumptions, excuses, unclear thinking, glib comments, Russian roulette, and false learning.***

Safety Mantra #12: ***What you don't know hurts and kills people.***

Take-Away Points: ***Chapter 12***

- Be wary of excuses and other indicators of hidden danger. Our excuses and our unsafe actions are transferred like a virus to our peers and live to infect future employees.
- No excuse can justify compromising safety.
- To get respect you have to give respect. Safety buy-in depends on mutual respect.
- Part of safety success is conveying to workers that they should always get the right tool for the job.
- People who say "Accidents happen" hold their mind in a state of deep unawareness and deny that accidents are avoidable, thus they display an ignorant resistance to learn.
- At-risk behavior must never be tolerated.
- The standard of proof for safety practices must include science and clear thinking. Don't believe everything you hear when searching for root causes. Clear thinking and resolving contradictions are necessary skills to overcome ignorance and assumptions.
- Unproductive fear holds you back. Facing each fear adds clarity and creates a path to the solution. This is your path of responsibility.
- Ask questions. Learn the stated safety policy. Find out who to contact if someone asks you to do something dangerous. Create alliances with other workers who will back you up. Create a plan before one is needed. Don't wait for a dangerous moment to make decisions. Be prepared up front.

- "False learning" in action creates unsafe habits and incorrect thinking, but does not change the statistical potential for an accident.
- Take those few extra moments to work the safe and correct way.

> *"If you don't find some way to discuss what's going on inside you, it can come out in other ways that are self-destructive."*
>
> - *Viggo Mortensen*

> *"If you don't think something can be done, at least get out of the way of the person doing it."*
>
> - *Robert J. Mathews, Former CEO of Cable Data and U.S. Computer Systems*

Chapter 13 – Examine Your Thinking Patterns

Achieving the next level of safety (zero accidents) is going to require you to examine your thinking patterns. Every person has thinking ruts - thoughts they think over and over again even if those thoughts have ceased to be relevant. It is another one of the weird things people do. Let's explain it this way.

Inertia

ANALOGY: Which takes more effort? Start a car moving from a dead stop or keep a car moving that is already rolling in the direction you want it to go?

Of course, it's easier to push a car that is already moving. It takes more effort to start the car rolling. That's because of inertia. Things that are "resting" tend to stay at rest. Moving objects tend to stay moving.

Cranes, for example, have large mass and a lot of inertia. A still crane loaded with iron beams will need a big push (a lot of energy) to start it moving - that's inertia. Once the crane is in motion and the load is swinging, it's hard to stop. That is also inertia.

Companies have inertia, too. And so do projects. And so do people.

EXAMPLE ONE: If a company has a tradition of bidding aggressively to win jobs, then that tradition builds inertia. It is easier for a company to continue to operate that way than it is for the company to try something new. And that momentum is strengthened each time frontline managers, supervisors and engineers are rewarded for designing jobs with cost-cutting perspectives and are acknowledged for bringing projects in on-time and on budget despite all odds. No matter what the stated safety policy, a company with production and cost-cutting momentum will look for ways to cut corners and set a priority on taking calculated risks and then hope they get lucky, in order to keep the "low bidder" momentum going.

EXAMPLE TWO: When a project's safety tone is based on policing and reprimands, the safety manager will be given great power. Workers learn to behave properly when the safety manager comes around because they know they will get in trouble if they are caught doing something wrong. This cultural mindset creates a certain type of inertia. It is easy to continue moving the project safely forward when the safety manager is present in the field. If the manager leaves, however, workers will let down their safety guard and accident-levels will rebound. This culture develops a dependency on the Project Safety Manager acting like a policeman, which is a dangerous development.

Now, these two examples point out the downside of inertia, but inertia is not inherently bad. In fact, here are some examples of good inertia:

- When a team has inertia, every worker understands the company priorities whether those priorities are announced or hidden.
- When a team has inertia, everyone from the company president to the daily worker knows what their top jobs are.
- When a team has inertia, the team works well together, every person knows their place and role.

So now we know that momentum can be good or bad depending on the situation and viewpoint. It is our job, as craft workers, to observe and monitor the situation on each site so we know what kind of "momentum" is on our work site and within our own work operations. Then we know what to watch out for.

For example, there was a project that hired many locals, but the project was frowned upon by the local community at large. On that project, the local workers kept getting hurt because the community's disapproval had "momentum" in their minds to the point where the safety manager could not leave the site during work hours or something bad would happen.

Frontline supervisors must keep constant track of the pulse of their crews. What is the momentum they are currently expressing? What subtle messages are being sent to the workers? What kind of example are you setting for your crews?

Sometimes, when psychological momentum is strong or has been around for a long time then people do not realize it is inertia and they think it is the truth. When inertia is so common it is below the level of conscious questioning, then these "truths" can become dangerous. Here is an example.

Ironworkers with a traditional daredevils-in-the-sky mentality and a love of autonomy, may find "all this fussy safety nonsense" inconvenient. It becomes very difficult to overcome their self-defining inertia when they have a "quick job" at a "not too dangerous height".

NOTE: Back in the early 1990's, Brion[55] *observed how Ironworkers resisted and often avoided 100% tie-off when doing elevated work. And the only project in Brion's long career where there was a fatality on the project was an Ironworker who fell to his death because he was NOT tied-off. Has the mindset changed today? Brion is not convinced. Oh, he will tell you that compliance has dramatically improved, but in large part it has to do with the threat of termination rather than fully embracing the use of life-saving PPE. This is intractable inertia.*

Imagine one person trying to stop a load of steel swinging from a crane when it is already moving towards them. It is impossible, right? Well, it is just as difficult to change your fellow worker's established mindset, tradition, or self-definition. It is especially difficult if that mindset is rewarded by management. But, the most difficult inertia to recognize and change is your own mindset.

Managing Tension

Here is a challenge. Imagine that you are at work and your co-workers and supervisor are standing around looking at you as you say the following. Check yourself internally as you say:

"Others don't believe I can improve safety to zero harm while increasing productivity, but I will stake my reputation on doing it."

Is there a tension in you as you realize these two parameters are in conflict? That's good because right now they *are* in each other's way. This is the correct tension to feel; get used to it. Learn it. Get comfortable feeling that specific tension. However, just because you are uncomfortable and do not yet know how to resolve that contradiction does not mean you are going to stop.

> *"Faith is taking the first step even when you don't see the whole staircase."*
>
> *- Martin Luther King Jr.*

The first step to making progress is to be able to see the difference between "what is" right now and "what you want" in the future. And you need to be consciously aware that your thoughts are in tension. We are going to tell you something very important.

Research shows that when a mindset is in tension, it can be dangerous. And it is very difficult to resist that tension.

This is very important. We will explain the science behind it in a future chapter, but realize right now that if your company tells you that you need to work quickly and

[55] Brion K. Hanks, one of the authors of this book.

you know you need to work safely, you are in tension - psychological tension. Your mind cannot sustain that tension for very long without constant and conscious support. If you do not consciously remind yourself over and over again, you will do something "unusual, strange, and bizarre" to relieve the tension and you will not even know why you did it.

Here is a quick analogy. Compare your psychological tension to a multiple-leg rigging. When you try to work faster while tying to work safely it is like the crane's angle gets steeper. More tension is created. This stress is dangerous for both the crane for your decision-making capability. If you do not account for the extra tension, then the crane could buckle under the stress. If you do not account for the extra tension, then you could do something dumb and unsafe to relieve the tension.

Psychological tension is one of the reasons people do weird things and cause accidents despite training, knowing better, and having great skills. Implied threats of termination make the tension worse. What you need, in this case, is to develop a way to look at the situation where the multiple sides of the tension are not fighting each other in your head and heart. It is such a simple answer that it is almost ridiculous. Great problem solvers are excellent at finding this common viewpoint.

Great problem-solvers (innovators) are experts at holding two conflicting realities *until...* they understand what they do not know yet and they continue to hold the two conflicting realities until they actually find the solutions that deliver the desired results or until the environment changes and the tension is relieved. You must eventually learn how to do this, too if you want to be a serious contributor to safety.

The hardest part of being safe every day on the job is going to be the internal work you must do and the shifts you make in your outlook. It is not going to be easy, but it is simple.

"Thinking is hard work. It's why so few people do it." - *Henry Ford*

Here are some of your first steps:

- You must know yourself – strengths and weaknesses.
- You must know that you cannot take a half-step across a canyon.
- You must have a BIG REASON to face-down a BIG TENSION.
- You must BELIEVE HARD in your BIG REASON.
- You must share this safety vision and be APPROACHABLE.
- You must put your faith in the tools we are giving you and in your talent and in your team.

Recognizing Negative Mental Inertia

So now you know that you have ruts or 'thinking patterns with momentum'. That momentum may be working for you or it may be hurting you. One of the most common ways it can hurt you is if you allow your thinking to put you in tension. If that tension is strong, you will be at great risk for making a blunder that could injure yourself or a co-worker.

Everyone has mindset momentum and it is usually useful but sometimes it is unhelpful. The question is: how do you know? How do you recognize when inertia is moving from a beneficial direction into a stagnant or destructive direction? The secret to recognizing this cross-over point is "problems". When problems suddenly start happening, your mindset has crossed over. Look at it from a project point-of-view.

When problems suddenly sprout-up all over the project, then there is an underlying thread of negative mental inertia in play.

- What to do: This is when you need to be extra careful. This is when you need to do some deep thinking about the underlying issues.

If parts of the project are failing or are behind schedule then stagnation is progressing. If the project is losing money and finger-pointing and blame abound, then you are at a cross-over point.

- What to do: Don't participate in the finger-pointing. Instead identify the negative momentum. It is probably due to the automatic response, "If I rush, I will save time." But we know that time will not be saved if you create an accident. Think about what processes can be improved. Start with the most dangerous processes in the critical time path.

If your problems have become extremely complex, pervasive, entrenched, or seem impossible then inertia is turning destructive;

- What to do: Watch out. Don't turn to compromises and trade-offs. Find a solution rather than a countermeasure for the long-term.

If solutions have been suggested and tried but are vigorously resisted, then mindset inertia is at the core.

- What to do: Discover what self-definitions are at play individually and in the group. Find out why there is resistance and the value inherent in it. Understand the hidden aspects of the situation. Use this information to create a robust and acceptable solution.

When your only options are to solve the problems with trade-offs that compromise safety then the situation is rife with psychological inertia.

- What to do: Know that this will put you mind in tension. You need to have constant vigilance and remind yourself to remain safe. Identify the contractions. Bring them out into the open and make sure they are acknowledged for the group. Reaffirm the commitment to use safe processes. Get the group involved in the solution.

From a safety perspective, you know you have unhelpful inertia when many incidents happen or "unusual, strange, or bizarre" accidents begin to occur. This is when people begin to do weird things. Problems are the way you recognize when inertia is moving from a beneficial direction into a destructive direction.

In addition, you can recognize that your own personal inertia is stagnating when you have lots of problems. Your mindset momentum is turning negative when you feel constricted or frustrated, when your options are narrowing, when disruptive forces loom and your influence is fading. When your future seems dangerous and conditions are worsening then your own personal inertia is unhelpful. When pressures are put on you because production is unsatisfactory and all that you try (or all that you know to try) is not working well then you are not addressing the real underlying problems.

If you do not easily fit within your organization or if demand for your services is lagging and you don't align easily with strategic goals then your own personal inertia is stagnating. If you are stagnating, you need to find a brand new way of thinking and you need to provide value in new ways.

All of the above is what the future holds for construction workers and frontline supervisors if you do not find new ways to add value in the future.

On a more positive note, you know your own and your project's inertia are still beneficial when:

- You and your organization are growing and vital.
- You arc compliant with your goals - always.
- There is abundant opportunity.
- The environment, situation, and conditions are stable.
- You and your team are productive and satisfied; morale is high.
- The workers and the whole project have a balance between stability and growth.
- You have plenty of resources now and for future needs.
- Precise feedback provides you all the information you need.

- Operations are aligned with strategy.
- Competitive pressures are easily managed.
- You have influence up and down the hierarchal ladder, can affect your environment, can affect your peers, and others look to you for answers.
- You can anticipate future conditions and know how to handle all disruptive forces that might appear.

Positive mental inertia means your grasp on reality is competitively strong. Detrimental psychological inertia means your model of reality is outdated, incorrect, and declining in relevance. Real world results help you decide if your mindset is helpful and relevant.

Thus, how we think (and act) creates our current situation because our choices proceed from our beliefs and knowledge. Just so, our current safety situation proceeds from our current beliefs, knowledge, and actions.

Let's walk through an example of identifying and overcoming psychological inertia. This example once again uses Ironworkers. The information is relevant for all jobs so watch your thoughts as we discuss this.

Falls are one of the most common causes of death for Ironworkers. Industry data shows that if a construction site is set-up with a strong emphasis on safety, then newly hired Ironworkers are not likely to fall. Data also shows that older Ironworkers (past their prime) are unlikely to fall. And statistics also show it is the Ironworker at the peak of their skill and confidence that falls most often. How can this be?

Could it be that it is not the level of skill that causes falls, but something else?

After much research, the conclusion that was hidden is now revealed. Complacency is deadly to the highly-skilled worker. We will discuss this in greater detail in the chapter detailing complacency.

Most craft workers would look at this odd trend and think, "That is weird" and then move on with their day. Not you. Going forward you must have a focus that is deeper than this surface thought. You must display more curiosity and want to learn more. You must also have the desire to recognize and overcome negative mental inertia when it develops if you want to keep yourself and your co-workers safe. Really think about this.

Psychological Inertia as Pain and Fear

Pain and fear form some of our most powerful beliefs and drive the way we think - pain and fear usually govern our choice-making. Break-through opportunities (like eliminating that last 5% of safety issues) are hidden within our pain and fear. We humans go to great lengths to avoid fear and pain. We will do more to avoid pain than we will do to increase our opportunities. Avoiding pain is a human necessity.

"Necessity is the mother of invention." - Plato, 360 BCE

Teams and construction worksite environments are developed by people doing whatever is necessary to avoid pain. If you look diligently, you will find opportunities that have been ignored, covered up, or had band-aid solutions created for them. Resolving pain and fear is a key to becoming relevant to your organization. This is a great career tip.

EXAMPLE: When we look around our workplace, we notice that some of our co-workers are better at sustaining attention for long periods of time than others. We discover that the ability to sustain attention is a useful skill. After further research, we believe people with this skill have different brain styles.

People with a particular brain style are better able to sustain attention on their work for extended periods than other brain styles. But wait! Don't jump to the conclusion that all construction workers should have the "sustained attention brain style".

Further research shows that people who are able to sustain strong focus usually hate to be interrupted when working. That makes sense - they dislike having their focus interrupted. This means they are very productive and produce quality work in protected areas, but are not so good in a part of the site where they have to keep track of what is going on around them as well as focus on their own operation. Also, since these people feel anger when their work is interrupted, if a foreman interrupts them they get irritated, which can be dangerous. If it happens often, then the worker begins to dismiss the foreman - and that could be very unproductive and can add risk. Also, if the worker needs to pick up a new pair of safety glasses because their glasses got scratched, that interruption would cause them psychological *pain* and they would resist doing it. But they will happily pick up all supplies before they begin their work. Now we have some interesting new insights that explain reality in more detail than before.

There has been plenty of research done on this subject of brain styles especially in other industries. For example, what kind of brain styles do neurosurgeons have to support performing six-hour operations? How about SWAT and hostage snipers? How do they sustain hours of extremely focused attention? What brain styles do they have? What are their stressors for losing focus? These are great questions.

While studying data is great, what is important for the daily worker is knowing how you run your brain when you are successful at your job and what is happening in your brain when you fail (whether you get caught or not). Most people are less practiced at examining their own beliefs and mindsets. If this is true in your case, you are leaving gains on the table and exposing yourself to unnecessary competitive risk, especially safety risks. Real leadership begins when you first challenge yourself.

Brion K. Hanks, one of the authors of this book, has heard about a very useful mechanics rule called the 20-20-20 rule. Every 20 minutes they take 20 seconds to look around within 20 feet of their work in order to confirm that there are no safety issues. This is a great way to break boredom, keep yourself aware, and refocus on the task at hand.

You must begin to do these things now if you want to be hired in the future when the industry is changing rapidly.

Groups, Psychological Inertia and Challenging Reality

Group thinking has a very powerful effect on our perception of reality. In fact, most of us humans are caught up in the mindset momentum of the community around us. You might have a mindset for the community your family interacts with and a mindset for your work life. Sometimes they are the same; sometimes they are not. You must get used to paying attention to the momentum of the groups in which you participate.

For example, group fear and pain produces powerful effects. The group's perception may become skewed as a result of avoiding pain and fear. Thus, one must be able to challenge the reality of a group as well.

Group negativity emerges and blossoms during poor leadership. The consequences are: key stakeholders become unresponsive, do not listen, or communicate ineffectively. This dramatically affects safety. Now that you know this, it is your responsibility to assume leadership when group negativity emerges. It does not matter where you are in a hierarchy, anyone can be a leader of a group mindset.

Group thinking can stink[56]. "Group Stink-Think" occurs when groups are operating in hidden fear and pain. This is when defense mechanisms are incorporated into normal group interactions and the group creates rules and truths that you must accept or be kicked out of the group.

This, in turn, creates an environment that actively avoids addressing issues of importance. There may be significant pressure to perform, but fear and lack of idea

[56] Group thinking can get blinky (not looking at things straight on and not looking for the truth), finky (mean and abusive), rinky-dinky (ineffective or deluded ideas prevail), shrinky (small-mindedness), overly sync-y (cliquish, exclusive) or let's call all of this 'stinky'.

diversity and stagnant processes often support stinky thinking in a group. Here are some characteristics of Group Stink-Think:

- People claim that their previous success justifies defining the world view for everyone;
- People have a narrow range of ideas regardless of the situation;
- The group has rules but those rules are not based on principles or reality;
- There is pressure to trade-off principles-based thinking for reality-based thinking or visa versa;
- People are detached or deluded about the current situation;
- People are using ineffective strategies;
- People are taking irrational actions or risks;
- People are trivializing the trends;
- "Insider" ideas have priority while "outsiders" have no say in their job processes;
- There is a strong them-versus-us mentality; dehumanizing others;
- Cliques-with-rules exist rather than cohesive and diverse teams;
- People criticize anything that would cause change;
- People tend to ignore alternatives;
- People have closed minds.

Any worker can challenge this "stinky" reality and counteract Group Stink-Think.

Expect some push-back, however. Persistence may be required and so tact will be required. Here are some measures to curb Group Stink-Think.

- Encourage a diversity of opinions,
- Eliminate threat and reduce fear,
- Convert fear into positive action,
- Openly solicit feedback-get people to speak up and then listen,
- Be open-minded and encourage open-minded thinking,
- Have a solution-seeking mindset,
- Use a facilitated structured process; sometimes an outside facilitator can break stagnated beliefs,
- Look for reality-based feedback on principle-based actions,
- Be observant and aware.

If you are trying to persuade a group to think differently or need to understand a group dynamic, you may need to understand who the respected leaders are (whether they have official authority or not). Determine their influence within the current situation. But proceed with caution. Do not allow your own biases and prejudices to derail doing what is best for the group. The winning formula is to realize that most of the time the people in a failing system can become positive contributors.

Carefully examine your own biases. Push yourself to consciously engage and sincerely listen to everyone, especially people with different opinions. Most people can become productive if they experience real leadership.

For those with influence but not authority, here is a very useful fact. If you "persuade" about a third of a population you will create a tipping point to adopt new behaviors and new solutions.

This is a large topic and we will not cover it in any more detail in this book. For our purposes here, realize that a strategic approach to influence involves challenging the reality of many people, but you can achieve progress one person at a time. If you want to be good at this, there is a whole discipline to study.

Bottom line #13: ***Examine your thinking patterns. Be responsible for your mindset inertia. Connect to your crew but be your own person standing up for the highest safety ideals.***

Safety Mantra #13: ***One for all, and all for one in safety.***

Take-Away Points: ***Chapter 13***

- Psychological inertia is not inherently good or bad. But if things are not working for us, then our current thinking is degrading in relevance.
- Companies have psychological inertia, too; and so do projects; and so do people.
- Internalized contradictions cause tension. Internal tension makes people blunder. Develop a way to look at the situation where the two sides are not fighting each other in your head, heart, and gut.
- The biggest source of tension and accidents on the jobsite is the safety versus production contradiction.
- Pain, fear and opportunity are the primary drivers of choice; but pain and fear drive us harder.
- Groups can drive us to think in similar ways, which can be good or bad. There are many ways to overcome collective stinky thinking.

- Group negativity blossoms during poor leadership, which harms safety. It is your responsibility to assume leadership when group negativity emerges. It does not matter where you are in a hierarchy, anyone can be a leader of a group's mindset.

"Between stimulus and response there is a gap, and in this gap lies all our freedom."

- Dr. Viktor Frankl

"Every mistake has a halfway moment, a split-second when it can be recalled and perhaps remedied."

- Pearl S. Buck

Chapter 14 – Outcomes of Split-Second Decisions

We have two stories to share that will illuminate two sides of split-second decision making.

A Risky Split-Second Decision

Many years ago Brion K. Hanks, one of the authors of this book, was visiting the San Francisco area. While there, he wanted to see the Golden Gate Bridge from a vantage point that was on the other side of a heavily traveled highway which had multiple lanes of traffic going in both directions. He stood and analyzed whether or not he could make his way across the highway on foot. Brion didn't want to have to drive away and find an exit where he could return to the other side. He couldn't believe how much traffic there was but decided to risk running across the highway. He was young, athletic, and single at the time.

The goal was across the highway - a great view of the bridge and the surrounding area - and Brion was determined to get there. He was focused on what he needed to do. He watched the traffic going back and forth and then made a split-second decision to dash across one direction of the traffic when there was a brief lull. He got to a narrow median. It became immediately obvious to Brion that he was standing dangerously close to fast moving traffic from both directions. Twice someone honked their horn excessively at Brion which was startling and reinforced the fact that he had put himself in a dangerous place. He is sure they probably thought that he was some drunken fool standing in the middle of a highway with nowhere to go.

Being pinned into place like that as a result of his decisions and actions was not a good feeling. Nearly five minutes went by before Brion saw a brief opportunity to race across the other lanes of traffic. He had made it to the other side but at what risk? When your heartbeat is elevated and you are in danger, adrenalin is pumping out in the fight or flight scenario. Brion freely admits that he was foolish to do this and he felt even more foolish that day when he later saw a pedestrian tunnel that passes under the highway which he would use to get back to the side where he had parked his vehicle after enjoying the view he had sought. Impulse can be a dangerous thing.

On the one hand, you could say that Brion only put himself at risk upon making that split-second decision; however, what if he had started to cross the highway and a concerned driver swerved or slammed on their brakes? Brion very well could have been the causative factor for a chain-reaction accident that resulted in other people being injured or killed. He was fortunate, lucky, call it what you will, that day because neither he nor anyone else got hurt. It was a lesson he never forgot.

And it is also worthy of note that the drivers that honked their horns to display their anger at Brion's presence in the median were also acts that were potentially life threatening to Brion and others if he had been so surprised as to fall or jump back and thereby create a cascade of reactions.

Brion's single split-second decision that day put many people at risk based on impulse. His desire for his convenience - to avoid driving past and returning by a different exit to get to the other side of the road - could have had disastrous results.

A Life Changing Split-Second Decision

One day Brion was having a personal conversation with a foreman he held in high regard. This man worked to improve project safety constantly. Their conversation started as a relaxed update on a home construction project but became very private as the foreman revealed intimate details of his personal life. It is not unusual for people to have this kind of connection with Brion as he makes it a practice to get to know people. This skill at connecting with people is a critical skill for a Safety Professional. (This story is masked for the sake of privacy while making the key point.)

The foreman had been working hard to make his house more livable for his wife. She was unhappy with the quality of their home and he was determined to make her happy. The foreman worked many hours at his regular "day job" in order to earn the means to pay for the remodeling project. He then spent nearly all of his spare time striving to get the job completed.

Brion asked if his wife was now happy with the remodeling. There was a pause, as if the foreman was deciding what to say to Brion, and then he answered, "I'm separated, headed toward divorce after 30 years of marriage, and the house is going back to the bank."

Brion's reaction was "Wow! How could such a dramatic change occur?"

Only a few weeks before, a co-worker told the foreman he'd heard through the grapevine that the foreman's wife had violated her marriage vows. "Thank goodness it was at the end of my shift, because I couldn't concentrate on anything else. I went home. I was in a complete rage." Up until that moment the foreman was self-described as a fighter and hell-raiser prone to a quick temper that sometimes got him into trouble. "I yelled when I got home and my wife was crying and scared of what I might do... Then in a split-second I sat down in my favorite chair and paused for a few seconds. I just stared at her."

"You sat down?" Brion asked.

"Yeah, the silence was deafening for both of us."

"I can imagine," Brion murmured.

"Brion, there was a part of me that wanted to kill her in that moment. I realized that I was about to make a decision that would affect the rest of my life. Suddenly, a sense of calm came over me. I gained complete control of my emotions and I decided to leave my wife rather than do serious physical harm or even worse." As the foreman looked into Brion's eyes, he conveyed how freeing that decision was. One moment he was hanging in the balance between right and wrong - between disaster and opportunity - and then, all the weight lifted.

The foreman then said, "There was a split-second where it could have gone a different way. In that moment when I decided not to be violent and the quiet calm came over me, I was a changed man from that moment on. Just like that I've moved on with my life and things are good in spite of what happened."

The future the foreman's life could have been one of constant regret. Instead, in a powerful life-defining moment, his friend chose a more enlightened option. Brion put his hand on the foreman's shoulder and they shared a profound moment.

Then, the foreman nodded his head and put on his ear protection - he was always safety conscious on the job - and he turned back to work. Brion smiled to himself, impressed as always how this man never let his preoccupation with this personal life affect his safety performance on the worksite.

Lessons Learned about Split-Second Decisions in Relation To Safety

You may be asking yourself, why are we sharing these two stories? What do they have to do with safety and risk on a construction job? The fact is that everyone has some measure of destructive behaviors and some measure of really outstanding behaviors in their personal lives. The decisions we make during non-work hours offer us important information about how we might act when stressed on the job.

From a safety standpoint, it is important to understand that there is a double-edged sword involving human behavior. It needs to be studied and discussed to really get to the bottom of why people make wise or unwise decisions that affect their own or other people's safety.

Sometimes the best people make really horrible and dangerous split-second decisions. And sometimes, the most unlikely people will make heroic, effective, and awe-inspiring decisions. We need to understand split-second decision making in greater detail so we can learn how to get people to make wise decisions in the workplace all the time.

The Double-Edged Sword of Human Nature

- One side of the sword involves human foibles and unconscious destructive behaviors
- The other side of the sword involves intelligent, wise, intuitive decision-making

People are complex by nature. We are each raised differently, make distinctive life choices, and learn through observed behaviors. These behaviors become embedded due to conditioning and usually we do not question them. Childhood experiences and intense life lessons mark us. Nurture, nature, and learning are all instrumental regarding the decisions we make each day, both in our private lives but particularly our decision-making processes in the workplace.

When our unconscious choices are positive, we call it our "intuition". When they are negative, we call it "destructive behavior". Some people are experts at making split-second decisions that turn out well. Other people's quick choices... uh, not so good. This is a constant and serious concern for anyone working on a construction site with other workers.

Can we make more reliable spilt-second decisions by combining intuition with logic, pattern recognition and science?

While this is an inventive question, it has been answered before in several other industries including: the military, medical emergency, the police, satellite orbit management, race car driving, air traffic control, emergency management, sports, and many others. People in these industries make decisions under pressure when time is highly constrained and the results matter. After careful research of these industries, we find that the main solution has been and remains "be prepared in advance".

> *"Prepare for the unknown by studying how others in the past have coped with the unforeseeable and the unpredictable." - General George S. Patton*

> *"You have to expect things of yourself before you can do them." - Michael Jordan*

For more reliable split-second decision-making in risky situations, the best strategy is to practice key skills through intensive simulations and scenario training. We have identified 10 key skills for split-second decision-making listed below:

1. Process Knowledge and Key Factor Identification

   In a rapidly evolving situation, there are some factors that matter more than others. These factors are critical to a specific situation. Also, to identify what is critical, a person must understand the basic processes involved in the scenario. They need to clearly understand how a situation evolves with all hidden triggers and patterns exposed.

2. Psychological Inertia Assessment

   Be able to rapidly and clearly distinguish assumptions from facts and the real from the unreal. This will save lives.

3. Constraint and Scenario Assessment

   Understanding constraints and limits can be valuable in an emergency. It is also valuable to rapidly evaluate a situation correctly.

4. Critical Thinking and Reality Check

   The ability to think clearly and ask tough critical questions can make reality much clearer.

5. Failure and Consequence Anticipation

   The ability to anticipate failure and see how the progression of events can lead to undesirable consequences is a critical skill set.

6. Self and Group Assessment

   Leaders and fools come in all sizes and appear at all levels of authority. Understanding and assessing yourself and group dynamics can improve decision-making under stress.

7. Body, Movement and Spatial Awareness

   An agile body, an understanding of space and the ability to move can help keep people safe. Not everyone has this skill and it can be vital knowledge in an emergency.

8. Intuition Training and Unconscious Risk Comfort Assessment

   It can be powerful to honor your gut feeling, the hair on the back of your neck, the unexplainable discomfort or inner voice that tells you what or what not to do in spite of all logic. You can learn to trust your unconscious wisdom and save lives. Likewise, one can assess their unconscious propensity to sabotage, to be clumsy or to take foolish risks.

9. Failure Pattern Recognition and Statistics

   There are patterns to how things break, how processes fail, and how people make mistakes. Learn these patterns, if you are serious; act like an insurance company. This is not for every craft worker. This would take exceptional effort.

10. Study Other Methodologies and Sciences

    This is also not for every craft worker; it would take exceptional effort.

Structured Innovation

The first methodology and applied science that we will briefly cover is called "Structured Innovation". It is a powerful set of skills and improves decision-making, cognition, critical thinking and a host of other valuable assessment skills. The more you use Structured Innovation, the more you train your brain to make concise, correct, "intuitive", and inventive decisions quickly. Structured Innovation also teaches people how to expose triggering beliefs and patterns and hidden priorities as well as how to think about problems so you can solve them for good (solution rather than a countermeasure).

HeartMath™

Another science-based factor is called "Heart-Based Scenario Evaluation" and is based on HeartMath. It has been discovered that the heart AND the brain work together to make decisions. Instinctively we know this because we say ordinary things like:

- "My heart just wasn't in it"
- "I don't have the heart to do that"
- "Trust your heart"

Over the years this scientific research has confirmed that group decision-making is also influenced by the hearts of the participants involved. HeartMath [57] is the science

[57] HeartMath™ is a registered trademark of the Institute of HeartMath™ www.heartmath.org

behind heart-based decision-making. And these revolutionary insights are backed by more than twenty years of research.

HeartMath has been used to help improve decision-making, health, intuition, symptoms of trauma, and Post Traumatic Stress Disorder (PTSD). It has been shown to reduce blood pressure and stress. While we have chosen to introduce this technology in this chapter, we believe it has broad applicability across the spectrum of improving safety, health, decision-making and lowering stress on the job. This technology has solid potential to improve safety and mitigate risk by becoming a common practice to enhance positive split-second decision-making in construction.

When it comes to making a split-second decision we feel craft workers can be more confident and centered if they combine intuition, logic, pattern recognition, and science with the heart's guidance.

Bottom line #14: *Split-second decisions happen in an instant but it takes intensive preparation to be ready for those moments where safety hangs in the balance.*

Safety Mantra #14: *Expect the unexpected and prepare for it now.*

Take-Away Points: *Chapter 14*

- Human foibles and unconscious destructive behaviors strongly harm safety. It requires insight, preparation, and training to overcome destructive behaviors.
- Intelligent, wise, intuitive decision-making can be enhanced through training and practice.
- Inventive Failure Analysis and Prediction can help identify situations where rapid decision-making is critical for positive safety outcomes. It is also useful for generating mental models and solutions for successfully handling stressful situations.
- Structured Innovation is a method for creating insights and breakthrough ideas on demand.
- HeartMath training is a promising new method to maintain calm under extreme decision-making pressure.

"It is more difficult to stay on top than to get there."

- Mia Hamm

"Concentrate all your thoughts on the work at hand. The sun's rays do not burn until brought to a focus."

- Alexander Graham Bell

Chapter 15 – The Complacency/High-Skill Contradiction

An interesting truth in construction: those at the peak of their skills are often the most susceptible to injury and harm.

It's hard to believe. It does not seem logical; however, we have seen this troubling trend in industry data. Because it is a confusing bit of data, it's critical to understand the pattern behind it. Getting to the bottom of this weird fact will teach us how to be safer for these types of situations and expose hidden beliefs in the larger safety system.

How can someone be at the peak of their skill and yet statistically be the most likely to have an "accident"? If you are a safety professional, have you been asked to explain this strange phenomenon? Why do very skillful people sometimes experience more safety incidents than apprentices or workers past their prime?

When we looked at this statistic, which holds true in other industries as well, we found a common root cause - complacency. During this research, we also found many different options on how to address complacency.

Complacency, as defined by Merriam-Webster's Online Dictionary, is "self-satisfaction accompanied by unawareness of actual dangers or deficiencies." The five stages of complacency[58] in the context of safety are:

1. Great Performance,

[58] Robert J. Latino, "*Distractive Environments: Mitigating Complacency*" HCPro, Inc. 2008; We compared our understanding of complacency that we created using inventive thinking (how to invent complacency) and searched for the best way to communicate this information. We used the 5 step process described in the above article to portray our similar conclusions. We also used Flow terminology and ideas by Mihaly Csikszentmihalyi and Brion Hank's safety data from on-the-job statistics to formulate an easy to understand five-step complacency process for safety in construction.

2. Self-Satisfaction,
3. Over-Confidence,
4. Misperception/Denial of Risk,
5. Safety Incident.

Complacency's Relationship to Unconscious Competence

The human nervous system and brain are amazing. Together they have the ability to create muscle memory, physical competence, and intelligence. When we learn and practice something, we generally become more competent at it. In fact, we go through four major phases of competency[59] starting with

- Unaware Incompetence – You don't know that you don't know - you are unconscious about it
- Realized Incompetence – You know you don't know and you realize you can't do it - you are conscious about it
- Alert Competence – You are aware that you know it and you can do it - you are conscious about it
- Reflexive Competence – You have high-skill that kicks in automatically and you don't even need to think about it - you do it unconsciously

This final phase is considered by many to be the ultimate form of competence, but it helps to understand what the brain is doing to achieve reflexive competence. Once you know that, you will understand why this can lead to complacency in routine dangerous situations (working on auto-pilot without situational awareness).

The human brain and the nervous system attempt to optimize its performance on an ongoing basis. The phrase "use it or lose it" describes how the human body makes decisions about what is valuable right now for your best interests. For example, if we are on a worksite next to a busy road, we probably tune out the sounds of traffic, horns honking, cars breaking, etc. But we pay attention to the warning alarms of forklifts backing up and the sound of our foreman's voice even though we hear them

[59] Noel Burch is credited with this work at Gordon Training International. He calls the four stages, unconscious incompentence, conscious incompentence, conscious compentence, unconscious compentence. We've only changed the words here in order to make it a little easier to understand.

often. The brain and nervous system make decisions about what is important for us to pay attention to and what we can tune out.

When we are trying to learn a new skill (welding, for example) we can tune out our awareness of the weather, or the tunes playing on a radio nearby, or the itch on our big toe. Instead, we focus our attention on what our mentor is saying and what happens to the metal as the torch tip approaches it. This ability to heighten and loosen focus helps us become competent at things we are trying to learn.

Learning is a high-brain-usage activity. To help us learn, the brain and nervous system take over many tasks such as breathing, balance, sight, and they even automate awareness when we practice something repeatedly.

When we achieve a high level of competence, we can perform the activities without thought - it is in our muscle memory. Also, when we become very competent, that competency usually becomes a part of our self-definition. This is when the trouble begins. Automatic skill can be the beginning of over-confidence.

Imagine this scenario: You have high-skill (are automatically competent) and you feel the job you are doing is not a challenge. You have no doubts about your skill and you have assessed the situation as ordinary. In this case, your brain automatically allocates less attention to this very routine (but still dangerous) task. In the five stages, this is the denial phase.

When you get really good at a task you let down your guard and you feel proud and self-congratulatory. Becoming cocky is just a few more steps away. Cocky means you do not define the risks as risky anymore and you feel you are better than the task at hand - that you are superior to the risk. With a gap between that incorrect perception and reality, one is now at risk of a safety incident. Think about it, should you really be driving a multi-ton vehicle while your brain is on cruise control. This above scenario is the recipe for a complacency-caused safety accident.

The Relationship of the Flow State to Complacency

Unlike automatic competence, there are other states we can achieve while working that do not lead to complacency. For example, we have all experienced a state of flow where we have pleasantly lost ourselves in a deeply engaging activity, but have heightened awareness of the task and the surrounding situation. Mihály Csíkszentmihályi, named this state "Flow" and it applies to working safely.

Flow theory states that three conditions must be met to achieve a flow state:

1. An activity must have structure, direction, and a clear set of goals.

2. Tasks must be clearly understood and feedback must be immediately delivered about performance so that performance can be adjusted to meet changing demands.

3. Your skill level and your perception of the challenge must be well matched. You must have confidence that you can effectively handle the tasks.

This is very good state of being for a dangerous job.

If we perceive our level of skill to be high and the level of challenge to be medium then we will have a feeling of control.

However, in the situation of over-confidence, our perception of the danger involved is less than the actual challenge and we perceive our attention needs to be medium. We feel we have more skill than required. There is a mismatch between the actual and our belief. Because the brain strives to shift us toward unconscious competence, this mismatch between real risk and perceived risk causes us to move into automatic action - we lose focus.

In the workplace, consciously and accurately gauging reality is better than allowing the mind to wander. If we deny there is a problem, then we are exposed to risk but do not realize it. This is dangerous and can have deadly consequences.

So from an inventive viewpoint, complacency is a measure of the mismatch between our perceptions of risk and the actual risk. It is also a measure of over-confidence or a loss of confidence.

Lying to Oneself - Another Weird Thing about Humans

There is an important cultural effect that makes it easier for a person to become complacent. The technical term is called the "Normalization of Deviance". It is basically how we justify lying to ourselves or accept our changing conditions without question or examination.

Here's an example. A construction worker is trained in the correct way to lift heavy objects. However, his right knee is a little sore, so for the following week he incorrectly picks up heavy bags of sand. Not once on Monday through Thursday does he hurt his back by working this way. He deviates (goes against) what is a normal best practice. By Friday, he develops an idea that lifting rules may not necessarily be true or they do not apply to him. Eventually, he thinks the safety best practice is too picky, and the standard way of lifting is not really necessary. He can lift in his own way without negative consequence. He accepts this wrong way of lifting as his new normal.

This creates and builds a momentum of false confidence. If he continues to lift this way, he will eventually damage his spine and soft tissues and he will not be able to lift heavy things at all. He has created a gentle path to accept a lie as normal.

The woman who identified this term was an investigator of the Space Shuttle Challenger disaster. She saw that this 'going against what is normal and forming false confidence' was a common effect.

In construction we see it, for example, when a worker fails to use safety glasses while drilling a hole or operating a grinder and no eye injury occurs. This "false success" reinforces the belief that it is safe to do so in the future, which exposes the worker to a statistical risk.

It is part of human nature to learn from experience. It is part of human nature to try to improve processes (keep trying different ways until the most efficient way is found). It is part of human nature to fall prey to the Normalization of Deviance unless high standards are held. The countermeasures for this are:

- Clear thinking, logic and science.
- Supervisors, perform field observations for complacent behaviors and mindsets especially during times of excellent performance.
- Train your crews to recognize the signs of complacency, which is: great performance mixed with self-satisfaction, a dose of boredom, a dash of denial of risks mixed with routine unchallenging work.
- Each person must examine their own psychological inertia especially during times of excellent performance.
- Reaffirm safety expectations and requirements.
- Reorient the complacent with the actual risks involved – a form of shock training.

Strategies to Eliminate Over-Confidence and Complacency

There are many strategies to conquer over-confidence and complacency. We will list one here.

The Value of Variation and Complexity

Research has shown that when variation is present, it makes paying attention easier because it keeps us on edge and therefore helps us to focus more attention to the task.

It has also been shown that lack of variation or very menial tasks reduces attention dramatically. In other words, if the task is too boring, the brain will automatically go into autopilot.

But, if the task is too complex, the brain goes into over-stimulation and it cannot pay close attention to that either.

To avoid danger, the task needs to be right-sized to your self-perception. Much research has gone into this topic of sustained attention - and variation and complexity are important ingredients.

After much research, the conclusion that was hidden from us is now revealed. Complacency is deadly to the highly-skilled worker because our brain puts itself on autopilot when we think the job before us does not need our concentrated focus or when we have done that kind of work before (no variation). When you are highly skilled, you will be unconsciously competent at many jobs and so you will be in danger of your brain "taking time-off" the job.

Add that specifically to an ironworker's traditional daredevils-in-the-sky mentality and a love of autonomy, and you have a serious potential problem on hand. So what can you do about this? Well, first, you need to get to know the people on your team and understand both their mindsets and how they feel about their own competencies. The Job Hazard Analysis is a good place for this – to openly discuss risks and consequences. Also if you are a highly-skilled person performing an easy task, you need to direct your full attention to prevent your brain from going into autopilot.

And each individual must take responsibility for their own mindsets. This problem with high-skill and low variation and complacency is now in your knowledge toolbox. You must manage that information about yourself continuously.

So now you know some of the reasons people do weird things, even when they know better. Unlike other craft workers who see these weird happenings and then move on with their day without examining it, you now can stop and understand what is happening. You can play a role in preventing issues from occurring. To add value in the future you must have the desire to recognize potential issues even before they develop.

Who Else Cannot Be Complacent

Who else faces severe consequences if they slide into complacency? What other industries or disciplines that are outside of construction are concerned with complacency? Here is our list of candidates:

- Healthcare in-hospital patient care and complex surgery,
- Military pilots and personnel in active combat,
- Air traffic controllers,
- Racing (cars, bikes, planes, boats...),
- High-wire acrobats.

We researched these industries and found they had many of the same problems as highly skilled construction workers. Many of these industries have developed methods for reducing complacency.

The top three best practices from these professions: Be in the moment, use your attention consciously, and use a checklist.

There are other industries that do not tolerate complacency on the job (working on autopilot). You can do a bit of thinking about who these individuals and industries might be. Then, do some research. Get to know high-skilled people in other industries and ask them what they do to keep their minds in the 'flow' and focused when their work is easy and repetitive. Ask how they keep from going into autopilot.

Going forward into the future, you must accept more responsibility for your work and your mindset. Do this research as an act of good faith that you will expand your current knowledge.

Creating Complacency - How to Be a Lucky Cocky Jerk and Get Away With It for a While

Because we are innovators we shift perspectives when we think we fully understand something. This new point-of-view helps us learn more about the subject. So now we understand that complacency is dangerous on the job. We also know how a person falls into complacency (high skill, low perception of risk). And finally, we know variation can help avoid it. To look at this differently, we are going to ask ourselves, "What are the steps I can take to create more dangerous complacency within myself?"

At first glance, we hope you are horrified with this suggestion. It sounds like a terrible idea. However, once we figure out 'the recipe' to create more dangerous complacency, new information and synergies are revealed, we develop new insights into the problem, and we create better tools to help counteract it.

Essentially, when we ask ourselves, "How can I create more dangerous complacency in myself," we turn on our creative thinking. This is a different part of the brain than the learning section of the brain. Using the creative part of our brains to "invent" a way to make a person dangerously complacent provides us with a process. Once we have a process, we can improve on it or interrupt it. We can use this information to keep highly-skilled workers safe.

EXAMPLE ONE: When we asked ourselves this question, we came up with the following recipe to create complacency. You need to:

- Be overconfident or fearless to a fault;
- Decide for yourself; disregard information that is different than your assessment;

- Be able to justify a drop in vigilance; keep your rationalizations to the forefront of your thinking;
- Be in a situation that is a mismatch for your skill (a low-risk assignment paired with a high-competency skill level);
- Believe that this task is "not difficult for me" (but the task is innately difficult or dangerous - or, it may be dangerous for other, less skilled, workers);
- Allow your mind to drift. Soften your focus. Notice distractions. Allow your brain to automatically move decision-making into subconscious mode;
- Decide that "I don't need to pay full attention."

If you have done all of these things, congratulations you are now a cocky, lucky jerk who is playing Russian roulette with your life and the lives of those around you. We're being a bit cheeky here to make a point.

Once you understand how these key resources create complacency, you can formulate strategies to interrupt the path towards complacency.

EXAMPLE TWO: Let's do this same thing again from a different point-of-view. During our research we discovered that coaches usually try to build up a person's self-worth, confidence, and belief in their skills. So, if we are tying to find out what to do to encourage over-confidence, maybe we can take what basketball coaches teach their athletes and push it a bit further.

- Practices dribbling so often that you don't have to think about what your hands are doing, only where you want the ball to be at any given time.
    - **On the job:** I am so good, I don't have to think about what I'm doing.
    - Do any of your peers or crew members think like this?
- Practice shooting from each spot on the court so you could do it with your eyes closed.
    - **On the job:** I am so good I could perform my work with my eyes closed.
    - Maybe it is okay to think this when you are off work; but never okay on the worksite.
- No pain, no gain.
    - **On the job:** Pushing myself to the breaking point makes me better.
    - Be aggressive. Be tough. Don't pay attention to limits.

- Be better than anyone. Don't let anyone steal the ball from you. Control the pacing. A little rough-play never hurt anyone, especially if you don't get caught - it's the nature of the game.
  - **On the job:** I'm better than anyone. No one is going to steal the work from me. I control the pacing. A little rough-play never hurt anyone, especially if I don't get caught. I'm the boss of this area.
- The guard's main job is to feed the ball to the Center.
  - **On the job:** Focus on your own job; let everyone else do theirs.
- Develop court-awareness. Pay attention to everything that is happening on the court.
  - **On the job:** Be aware of everything on the jobsite.
  - This is the opposite of the one directly above. It promotes distraction.

What we can see from Example 2 is that good behaviors in one environment (coaching young basketball players) may actually contribute to unsafe work practices in another situation (on a construction project). Sometimes useful information has a context. Information is not good or bad, it is only the situation that makes it so. Context (the situation) is vital when considering information.

EXAMPLE THREE: Here's good coaching advice for a person trying to move up the corporate ladder to build their self-confidence and people skills; but these attributes could be ominous when possessed by construction workers who perform dangerous work. These habits and traits may promote complacency, loss of focus, or over-confidence. Can you figure out why these would be bad in a construction site (especially if pushed too far)?

Expect the best from yourself and others. Pay attention to and enjoy the successes of your peers. Smile and acknowledge everyone you meet.

- **Good coaching advice but pushed too far for workplace safety:** if you are focused too much on how well other workers are performing, you are not focused enough on your work or safety.

Develop a peace of mind so strong nothing can disturb it.

- **Good coaching advice but pushed too far for workplace safety:** If a dangerous situation is developing, you want to notice it.

Let go of the past, especially forget and forgive your past mistakes. Focus on the future.

- **Good coaching advice but pushed too far for workplace safety:** Every mistake needs to be a lesson so it is not repeated. The focus needs to be on now and what is unfolding around you.

Set yourself above worry and fear. Do not let troubles affect you.

- **Good coaching advice but pushed too far for workplace safety:** This is a recipe for complacency and denial. You need to worry just a bit to stay safe.

Can you see how each bit of the career coach's advice could lead to disaster? Safe projects want attentive craft workers. This is true. Safe projects want confident craft workers. This is also true. But don't we also want highly-skilled craft workers to be just "nervous" enough so they stay focused on their work?

Perhaps the safest and most productive craft workers possess the skill to sustain attention or shift states as the situation warrants.

This is a great insight and deserves some thought. How can you develop the skill to shift states from 'focused and attentive' to 'widely aware of your surroundings and cautious' and then back to focused again?

Great construction workers need to be able to move into and out of degrees of focus; widening and narrowing their attention as needed; watching out for peers and focusing in on their own work...

So if you are highly-skilled, it may be better to increase your sense of unease, keep your mind focused on each careful step, or keep yourself from feeling too competent or confident in your job. Maybe you can find a way to compel your attentiveness. How would you keep your highly competent minds fully concentrating when it is the brain's natural tendency to move common actions to the unfocused area of the brain?

So now we understand many aspects of complacency. We have also identified some basic ways complacency happens and now you can make sure those "bad" resources are not available for use by you or your team when you are performing your work. We have found some areas of research you could perform if you wanted to take this further. And we have stressed the importance of taking responsibility for your mindset. And we have also given you ways to assess when your crew's mindsets (or your peer's mindsets) could be harmful - especially if pushed too far.

Here is the bottom line: Complacency must be eliminated on the job.

The Power to Choose

The power to choose one's feelings is an important ingredient in recognizing danger and paying attention. In fact, it's huge in the scheme of a craft worker's daily work activities. Vigilance and paying attention are present-moment activities. To have excellent control of one's mind, you must

- Consciously choose safety and awareness moment-to-moment.
- Be able to get into the flow of the environment and the work at the same time.
- Be able to shift from very-focused to very-aware.

Meditation helps train the mind. Sports help many people train their minds. The martial arts are very good at this type of training. There are other methods as well. Perhaps you can ask around to find a way to train your mind that you would enjoy.

Choice is fundamental to safety success and workers must choose to implement right action in order to achieve safety perfection. Supervisors have an extra responsibility. You must understand each member of your crew. You must understand which aspects of their mindset and habits are useful and how they may cause safety issues. You must figure out a way to train your workers to choose safety moment-to-moment and to train your workers to be able to shift their focus and attention as needed.

Here's a way to practice training your mind for workplace safety: While driving, while eating dinner, or while doing any activity that you tend to fall into "auto pilot", push the pause button in your mind. Allow yourself time to analyze the situation. Make a focused decision to check the mirrors or savor the taste... What does it take for you to stay focused and engaged throughout the whole drive home or throughout the whole meal? Can you recognize the triggers that cause you to fall out of focus? What do you think, feel, and say to yourself? Are there environmental triggers like slow music on the radio lulling you? Does a rash of kids all talking at once over stimulate you at the table? Practice paying attention outside work to help yourself remain safe during work.

Choose to act courageously, pay attention, and speak up about this safety issue to your peers. You may save a co-worker's life. To avoid complacency, consciously choose to be attentive and safe every moment, during every operation on every project.

Situational Relevance

Having a focus on the changing situations and recognizing what you need to be or do to meet that situation is not only important, it's vital. It would be useful to train workers to make better split-second decisions. This is something that you could ask your company to provide if you think they would be open to these kinds of suggestions. This kind of training is described in Chapter 14.

The Value of Arousal

Arousal is a scientific term. Research has shown that a person who is "aroused" is emotionally engaged and stimulated and they will perform with constant awareness and diligence. There has been a lot of research on the topics of arousal and performance in other industries.

When a person is in the state of arousal they perceive that there is risk and a need for attention. When aroused, you are able to focus in on what you are doing and at the same time you are aware of what else is going on around you. For example, you are open to all feedback. You are aware of subtle feedback - like a shifting facial expression - and are aware of primary feedback such as a mentor giving a demonstration.

An example of this arousal state is 'a growth experience'. Growth experiences are when your skill is barely adequate (you can do the work but you do not have full confidence) and you want to perform well. During this situation you are stimulated, diligent, and very aware of what you are doing and you are also simultaneously listening to your mentor's feedback.

A "proper degree of risk", therefore, stimulates attention and arouses focus, which generally leads to safer job performance. This is a topic you could do some research on or practice if you wanted to increase your ability to enter aroused states.

Congruence of Knowledge and Action

Part of complacency is not acting on what you know. Training and culture can help with this. When there is a gap between what you know and what you are doing, in that moment you have likely become complacent and are moving toward apathy but you still believe you are acting skillfully.

As a supervisor, if you received training to recognize this condition in others, you would be a valuable addition to any project. Your ability to keep yourself and your co-workers safe would be noticed over the long term.

You need to learn to recognize the states of arousal or boredom and complacency. Become competent at this by observing yourself first, and then you can apply your trained eye to the activities of individual workers and crew dynamics. This is a true sign of leadership.

Optimal Engagement of Automation and People

Most people fear automation because it often eliminates jobs. However, automation of work holds high potential to keep people engaged with their work and in the flow.

Although Csíkszentmihályi asserts that the Flow State just happens and cannot be forced, we know otherwise. The human brain and nervous system have processes to become engaged in the flow state. We have long since realized that the only reason "you cannot force a flow state" is the process drivers for a flow state are not fully understood. All processes are the result of key factors that generate outcomes of the process – in this case a "flow state". We simply need to discover these key factors and harness them. We expect that the future of construction will use automation to cause you to be more attentive and not to let the automation allow you to fall into auto-pilot.

If you are a naturally observant and process-oriented person, you can study these conditions in yourself and others to learn how you produce a flow state and enjoyable concentration whenever you need it. We think that meditation is a very similar process to the flow state.

Keep your ears and eyes open for industry breakthroughs in the science of automated work that adapts to an individual by allocating work dynamically.

Research shows that optimal engagement with automation involves creating meaningful work to keep people's attention. That means as work becomes more automated (and/or repetitive) it must also be enhanced to engage an individual's arousal strategies.

You can create this type of enjoyable concentration now without the automation and make yourself and your peers safer.

The Value of Hypnosis and Positive Self-Talk

Language can have a strange and powerful influence on people's behavior. Language combined with suggestion and hypnosis has accomplished positive results for top athletes. It helps them get their mindset and focus in the right place. So why not use it for improving safety and especially improving attention, focus, and to help workers accurately gauge risk?

Hypnosis has been shown to reduce stress, improve weight loss, and have strong physiological effects on the body related to perception of pain. Changes actually occur in the brain during hypnosis verifying the effect is real. It has also been shown to be more effective if used over time.

We feel this is an underutilized opportunity and a practical element of improving workplace safety and eliminating complacency. Many consider that hypnosis is not real, but the science in support of it is overwhelming. Have an open mind about this. Remember, the object is to eliminate accidents that injure or kill people. A tool not tried or researched is a tool wasted. But remember, a strong ethical approach with independent oversight is strongly recommended.

HeartMath Process

We discussed this research previously in another chapter (Outcomes of Split-Second Decisions), but it is worth mentioning again. Sustained attention, vigilance, focus, and stress reduction can all be accomplished using HeartMath training.

The scientific results derived over 20 years have shown that these methods can improve performance in high stress jobs like in the military or with police personnel.

We feel that it can be applied to workplace safety very effectively. There is also evidence that it can be enhanced when combined with hypnotism.

Conclusion

Understanding complacency is critical to improving safety for highly-skilled craft people. And if complacency is reduced, safety will be improved for all craft workers.

Automating certain activities on the worksite could make the work easier and less dangerous, but it may also lead to complacency because it reduces the complexity of a job. It is the responsibility of the craft worker to develop a strong flow state and meaningful engagement with their work.

Bottom line #15: ***Reduce complacency by keeping yourself engaged and attentive to the task at hand.***

Safety Mantra #15: ***Safe work demands sustained attention.***

Take-Away Points: ***Chapter 15***

- Workers at the peak of their skill are often the most susceptible to complacency.
- Complacency happens because our brain puts itself on autopilot when we think the job before us does not need our concentrated focus.
- The brain and nervous system make decisions about what is important for us to pay attention to and what to tune out based on "use it or lose it". This ability to heighten and loosen focus helps us learn.
- When we are competent, we can perform the activities without thought - it is in our muscle memory. Automatic skill can be the beginning of over-confidence.
- Not all high-competence states are dangerous. Flow states are deeply engaging, highly vigilant, and adaptable.
- Normalization of Deviance is a process of the gradual acceptance of lies. These lies kill and harm.

- Complacency may be managed with the right amount of variation – the job should not be too complex or too simple.
- The safest and most productive craft workers may possess the skill to maintain or shift states as the situation warrants.
- It is within your power to choose how you will run your brain. Be vigilant, pay attention, and train your mind.
- Hypnosis, HeartMath and other tools can train your brain and nervous system to be highly vigilant and avoid complacency.

"When people restrain themselves out of fear, their lives are by necessity diminished. Only through freely chosen discipline can life be enjoyed and still kept within the bounds of reason."

- Mihály Csíkszentmihályi

"Situational variables can exert powerful influences over human behavior, more so than we recognize or acknowledge."

- Philip G. Zimbardo

Chapter 16 – An Inventive View of a Culture of Safety

In Part II we discussed the current safety culture in the construction industry - the best practices of what we currently know as an industry. By now we all agree that it is not good enough, but it's the best we can get from our current knowledge toolbox.

We also know that one of the biggest obstacles to a perfect safety record is people. People do weird things and they don't always know why they do them. We've discussed hidden triggers and hidden patterns that cause unsafe behaviors. In this chapter we will present additional information and add new meaning so we can add more tools to our knowledge toolbox and help keep people safer on the job. Some of our new viewpoint will be viewed through the lens of teams.

NOTE: We will focus on a vital few topics that matter the most so you can begin to improve your own culture of safety no matter how robust it currently is. But understand this: science will continue to advance and it will be up to you to keep learning throughout your whole career.

More Strange Things about People

In this section, we look at the weak points in human nature and thus safety culture. We examine individuals and groups through the lens of a darker side of human psychology. We will reveal what we believe are fundamental interfering factors to reaching a 100% zero harm culture of safety.

Brion K. Hanks, one of the authors of this book, has observed throughout his many years as a safety professional that people often do not accept the responsibility for safety on the job fully into their heart. While this does not usually affect day-to-day safety, it does occasionally result in extraordinary events that end in death or injury. You may recall some of the strange behaviors we reviewed in the previous infamous historical accidents.

There are master patterns at work in human psychology and behavior that lead to risk. The primary negative factors that can lead to strange, dramatic and harmful events are surprising:

- Lying to yourself (Normalization of Deviance)
- Foolish obedience,
- Fear of being wrong or criticized,
- Situational uncertainty,
- Ambiguous social situations,
- Negative social support,
- Negative mindsets,
- Knowing better but doing it anyway,
- Inaction and diffusion of responsibility.

Let's get cozy with these issues so that you can understand how completely rational people can behave in ways that are irrational and bring harm to themselves and others. Then we will discuss how to mitigate or eliminate these issues.

Lying to Yourself (The Normalization of Deviance)

We discussed this in Part II, but it deserves mention again.

Everyone is susceptible to this phenomenon. It is liking the "boiling frog" scenario.[60] Over time some practices become normal even if they are not safe because we accept small lies without examination over time. Below are examples of small lies that grow over time:

- Driving a truck without a certification is okay even though the evidence shows that accidents occurred from uncertified drivers.
- The Space Shuttle Challenger disaster occurred because it became normal to analyze burning O-rings to confirm they will not fail even though they separated oxygen from hydrogen - an explosive combination.

[60] The Boiling Frog scenario is an anecdote that explains the normalization of deviance. Here is the metaphor: if you throw a frog into a pot of boiling water, it jumps out. If you put a frog in a pot of cold water and heat it gradually, it will stay until it boils to death. Because the changes are small the frog ignores the changes and accepts them - the frog normalizes the heat. This is the same as lies we tell ourselves. The normalization of deviance can kill us.

- It is normal to climb those beams without a tie-off because my grandfather built the SF-Oakland Bay Bridge that way. He died by falling from that bridge but that is the risk you take if you are an Ironworker. Minimal safety gear, few regulations, and just a little bit of crazy (said with a smile and tacit approval) gets it built.
- It is normal to put people at risk because I like to do things my own way and it gets done faster.
- And it is okay for people in construction to get hurt or die on the job because civil infrastructure is so important and saves lives.

Wake up, people! Really??! Look and see what you actually believe! When we start out afraid we try to make it noble to take risks. This is a countermeasure. We say it is inevitable that some will die and then we turn off our brain.

Instead consider this scenario:

1. Hold yourself to high standards for safety.
2. Think clearly and face your fears.
3. Consider it noble to accept the responsibility to set high goals and solve all sides of the challenge even when no one else thinks it can be done!

Go for everything you want not just some of it! This is how you eliminate The Normalization of Deviance – with greatness.

Foolish Obedience

In a landmark experiment in 1961, Stanley Milgram asked participants under study to push a button and deliver an electrical shock in increasing levels of strength when the actor (involved in the study) missed an answer. An authority figure in a white lab coat stood near the button-pushing participant. No actual shocks were being delivered but the actor behaved as if they were being shocked and the button pusher believed the shocks were real. The participants under study (button pushers) were paid to be in the study.

After the actor had missed several answers (on purpose) the electrical shock levels were "increased" to dangerous levels. The participants continued to deliver shocks even though the actor was complaining that it hurt and was yelling. Many of the participants requested to stop but when urged to continue by the authority figure in the white coat, they did continue to push the button and deliver the pain even though they clearly did not feel comfortable. Some participants continued to deliver shocks, even when they believed it was a lethal dose, because the authority figure said:

- "Please just continue."

- "The experiment requires that you go on."
- "It is absolutely essential that you continue."
- "You have no other choice, you must go on."[61]

In Milgram's experiment, 65% of the participants delivered what they believed to be a lethal dose of electrical shock.

Looking back at the chapter of accidents in Part II, do you see a pattern here? It is very hard for most people to say "No" to authority even when it is clear that an unsafe condition is present at a lethal level. They are deferring their authority to someone else.

Also, they do things they know are wrong out of a sense of obligation.

Also, the people believed in the power and value of science - actions performed for the greater good. Construction workers also believe in the benefits of their work. In the back of their mind is the idea that acting unsafe (just this one time) is for the greater good. This is a big contributing factor for accidents. We need to fix this if we want a safer jobsite.

Obeying authority is necessary in most of our daily decisions, but for a truly safe environment one must remember obedience is a CHOICE. It is good to obey when it makes people safe. But workers must constantly communicate with each other, walk the talk, participate fully in training, and give their co-workers positive reinforcement that it is okay to say "No!" to unsafe acts and to personally deal with or support correcting an unsafe act.

What about the 35% who did not go on shocking the person? What did it take for them to choose to stop even when an authority figure told them to continue? What kind of character or self identification allowed them to follow the rules until... and then they made a better choice? What would you have to develop in your own character?

It starts with each worker being a living example of standing firm. When your peers see how you value the true principles of safety by standing your ground against dangerous demands, you become very powerful and very productive. Co-workers need to know that you will not commit or participate in dangerous activities no

[61] We reviewed multiple sources for this information, but the most compelling is the actual video of the study while it was happening. *The Milgram Experiment* (Full edited Video); just under 16 minutes long; Presented by Michael Portillo, Narrated by Paul MGann and available on YouTube. We accessed it at http://www.youtube.com/watch?v=pdb20gcc_Ns (last accessed 31august2013).

matter what the provocation or take any action that puts others at risk. You must become a superb safety asset at the same time you are productive and efficient. It can be done and must be done.

Fear of Being Wrong or Criticized

Another master pattern of human psychology that leads to risky behavior is "Fear of being wrong or criticized". Sometimes it can be hard to ask a question for fear of criticism. We've all worked with people who are more than happy to let you know when you are wrong and most people have made mistakes and wanted to avoid that criticism.

When you are in a great company, you notice the safety professionals are very good at limiting this negative talk. And yet, they are also very good at engaging people to speak up. This is a contradiction. You need people to be vocal and supportive of safety programs AND you need them to speak up and provide inputs that are sometimes very negative. What d you do?

You develop guidelines for the "right way of being negative". Here is a good guideline for how to be precisely negative:

- Focus on issues and roles and not specifically on people.
- Be precise, factual and exactly describe the negative issue with the intent that someone is going to problem-solve.
- Complaining is not problem-solving. Complaining is a distraction from solving problems or presenting grounded data. Instead provide supporting information. Discuss the scope of a risk. Explain how it could happen again. Work with the other person to make it better.
- Discuss the situation in order to define the problem or contradiction and, if you can, make it a simple contradiction so the problem will be easily solved.
- Use neutral language to minimize emotion.
- If a negative person is a natural leader and influencer the supervisor will likely need to have a serious private discussion.
- When speaking to a group, only involve names if a person is actively reckless, drunk or on drugs (and only when the private discussion failed). If a person's name must be used publicly, be considerate and careful in how you do it. Always inform a person in advance if you are going to use their name publicly.
- Do not allow general negativity and imprecise sweeping statements that cannot be verified.

Drive fear out of the organization and leverage precise negativity to create mutual trust and a problem-solving team spirit. It can be done. It needs to be done.

Ambiguity and Uncertainty

When it is unclear what is happening, people will often make poor decisions. Ambiguity and uncertainty influence how people will behave in group situations. This often leads to dangerous results.

In one very strange example, a group of people did not exit a burning restaurant because they were waiting to pay for their meal before they left. They were overcome by smoke inhalation and died. They stood in line even after others had fallen. This type of behavior has been seen repeatedly in major emergencies when it was hard to accept that the reality was actually happening.

During the rise of Nazi power in Germany, many Jewish people were systematically killed or removed from society without "official notice". It was very difficult for the soon-to-be victims to accept that it was actually happening until, in some cases, it was too late to escape.

There are many documented cases of people that refuse to evacuate when a flood, hurricane, volcanic eruption or other major emergency is present. Even when authorities are adamant that the danger is real and present, some will not accept the reality.

Even when facing death by choking in a public restaurant, after their rescue, the victims admitted that they did not want to draw attention to their condition because they felt embarrassed they were choking.

The ability to accurately assess reality, determine with confidence that danger is present and to act calmly in the face of the danger is a critical skill and valuable countermeasure. And finally, getting others to overcome their social obligations, acknowledge danger is present, and get them moving in a calm and reasonable manner is real safety leadership.

Negative Social Support and Negative Mindsets

People will behave differently depending on the social forces. Peer pressure is a well-known and powerful force. We usually think of peer pressure as only applying to teenagers. But the lessons of Nazi techniques shows that ordinary people will do heinous things and not even flinch if they are told to do it by a strong authority and have social support backing the orders. Here are common negative forces involved in getting moral people to go against their integrity and feel it is routine:

- Belief you are facing a fully committed or irrational "enemy",
- Strong "us" vs. "them" mentality,

- The enemy is branded and deprecated by leaders,
- Management and peer use of slurs, disgust and deprecation of "them"; social rewards are given for mocking "them",
- Severe peer pressure and structured social conformance,
- Urgent goal pressure related to a "misbehaving them",
- Critical need for immediate results,
- Vague orders with high expectations of performance and minimal oversight,
- Implicit rewards for delivering results at any cost,
- Leader or social "weasel wording" (creating euphemisms) and degrading the meaning of commonly understood human standards,
- Secrecy and shrouded actions,
- Massive social support,
- Uniform dress code, slogans, behaviors and rules for "us",
- Insider rules and methods for handling "them".

These dynamics encourage abuse and stifle "speaking-up". These factors occurred in full or in part throughout WWII Nazi occupied territories, during college student hazing incidents, at Abu Ghraib Prison, and during the Stanford Prison Experiment[62]. Soberly consider the comments from the historical leaders below.

> *"Let us not forget who we are." - Ronald Reagan*

> *"If words of command are not clear and distinct, if orders are not thoroughly understood, the general is to blame. But if his orders are clear, and the soldiers nevertheless disobey, then it is the fault of their officers." - Sun Zu, "The Art of War"*

Phil Zimbardo, retired Stanford University Professor, is famous for his Prison Experiment. Zimbardo summarizes some of the forces involved in helping people choose to behave badly:

- One group was in authority over another in a Them-vs.-Us separation,

62 Although there are many papers and analysis of this experiment, there is a compelling video on YouTube that summarizes it nicely. *The Stanford Prison Experiment*; Narrator David Suchet, Archive: Philip G. Zimbardo, Inc.; Obedience: Stanley Milgram; CBS News; Produced and Directed by Kim Duke, BBC located at http://www.youtube.com/watch?v=760lwYmpXbc (last accessed 31august2013).

- Rules were imposed on the prisoners (Them),
- Strong Social Support for the guards (Us),
- Secrecy and shrouded actions,
- Guards (Us) wore uniforms and dark glasses and enjoyed delegated authority as well as a great deal of autonomy (over Them).

With only a handful of the major forces listed previously, the experiment careened out of control with the subsequent breakdown of normal standards of behavior and systematic abuse began to take hold. They were forced to shut down the experiment early. Phil Zimbardo is now running the Heroic Imagination Project which seeks to understand these forces and what makes a person an everyday hero.

Now look at the same list through the lens of safety in your organization. Do these exist in your organization?

- Belief that safety professionals and OSHA inspectors are jerks or are an irrational self-serving "enemy"?
- Strong "us" vs. "them" mentality (against other crews, subcontractors, etc.),
- Safety professionals are branded and deprecated by leaders or peers,
- Management and peers use slurs, disgust and deprecation of "safety professionals". Social rewards are given for mocking "them",
- Severe management and peer pressure to deliver on time,
- Critical need for immediate results,
- Vague orders (minimal process controls) with high expectations of performance and minimal oversight[63],
- Implicit rewards for delivering results at any cost,
- Management or social "weasel wording" and degrading the meaning of commonly understood safety standards,
- Secrecy and shrouded actions (behaviors are different in the presence of a safety professional),
- Massive social support,
- Uniform dress code, slogans, behaviors and rules for "us",

---

[63] 'Looking the other way' is one example of a minimal process control scenario. Reading the accounts of the situation during the construction phase of the City Center Projects in Las Vegas, an observer could conclude that this was a giant "Milgram's experiment" played out in real life. Hmm?

- Insider methods and rules for handling "them" (for example, OSHA, subcontractors, or safety professionals).

Take a moment and breathe. Did you learn something valuable from experiencing culture in its negative forms? You may feel elated because you have faced such issues successfully or you may feel down because you see yourself in some of the negative behaviors. Be kind to yourself and realize this is why it is so hard to overcome the last 5% and why it is not valuable to judge others but to be a part of the solution. You must have a strong heart and head to be a safety leader.

What does your culture allow in terms of negative mindsets and negative social support that allows the jobsite to become unsafe? Is hazing tolerated? Are negative "leaders" allowed to bully others on the job? Are company employees allowed to separate themselves from the subcontractors and enjoy greater privilege, and better equipment, etc.? Do people fear their supervisor? Are workers afraid they will be outcast for speaking up against pervasive unsafe behaviors?

If these attitudes are present in your organization then you have many safety time-bombs ticking which are ready to blow at any moment. Every single issue on the above list can be effectively addressed.

To begin with, leaders at any level must be effective at standing up against such forces. What would it take for an individual to go against the momentum of an entire project site population, management, or even the entire organization? What kind of integrity must that person have to expose abuse or dangerous policies? It takes character to stand up to multi-tiered authority when they say, "We're behind schedule. Don't take the time to put up that safety tape. I don't think anyone is going to be walking near that hole anyway." What kind of leadership does it take to report clear evidence of safety violations that are becoming systematic and dangerously out of control - and that are damaging to good employees' psychological health?

It takes strength of character to stand up for what is right, whatever the level of authority. That person would be a valuable asset to making the workplace safer. Such a person would have confidence and a strong sense of ethics. They would fit into the team, but be independent of social group pressures. A person like this chooses to be a part of the group, but does not automatically conform. Leaders, especially safety leaders, must act on principles and less upon social pressures. They must be more integrity-based than social-based but they must be able to gain rapport with others (up and down the hierarchical ladder) to effect positive safety changes.

Strong leadership and character can overcome negative social support and negative mindsets by breaking the social conditions that allow abuse to flourish as previously discussed.

Negative Psychological Inertia

If a problem is longstanding, it has negative momentum (inertia). If a problem is entrenched, traditional, pervasive, or involves many issues then it has negative inertia. Here are some of the red flags of negative inertia.

The Realm of "I Can't":

> *I can't do it. You can't do that. That's impossible. Nobody can do that. You are going to fail. You're asking too much. Are you kidding?*

Negative Claims:

> *I am not that kind of person. I can't change. I am not the learning type. You can't teach an old dog new tricks. I am too old for this. That's bull**t. That is someone else's job. It is not worth it to try. You're going to fail if you try that.*

Rationalizations and Avoidance:

> *Management wants it done now. We'll be okay this once. I have my job to think about. Let's deal with this later. Let's not get ahead of ourselves. You are always thinking. Shut up and get it done. No one cares what you think.*

These three forces create an expectation of failure before any effort has begun. Negative psychological inertia is so powerful that even great innovators and leaders can experience an altered reality.

You must discern if you are trapped in a false negative reality or if you are creating reality with your negativity. Take a moment to really digest this.

The Remedies:

- Perform a reality check and use critical thinking.
- Say, "If I can't, then I must, so I will."
- Think, "If it is to be, then it is up to me."
- Prioritize, "Later means never, so it must be handled now."
- Make your safety motivation strong, passionate and obvious.
    - A great example that we previously mentioned is Kiewit's Rand Magee who wears a picture of his family on his safety vest so everyone can see at a glance why he works safe. This is a powerful subtext.

Knowing Better but Doing It Anyway

Why? Why? Why do people know better but do the wrong thing anyway? This is such a basic problem in the construction industry and is so entrenched it seems impossible to overcome.

Great problem-solvers, however, do not stop at impossible problems. In fact, these are exactly the kind of challenges that make our blood stir and excite our curiosity. So, after some thinking and research, we are able to pull back the curtain for you and give you a peek at a potential solution.

Why do good people, who know better, take wrong actions? Because: our brains are hardwired to do this.

"What???" We hear screams of anguish from craft workers and safety managers all over the world. Are you saying there is nothing we can do about this problem? No, we are not saying that. In fact, you are doing a lot of the correct things right now if you are using the best practices outlined in this book. And, you can do more immediately if you just understood how we humans are hardwired. And there are even more benefits you could reap if you put time and attention into it. Let's get that curtain out of your way.

Leon Festinger created the Cognitive Dissonance Theory, which basically states that humans have an abiding need to be consistent in our attitudes, beliefs, and behaviors. We must feel in agreement with our thoughts, feelings and actions. We need to be unified within ourselves. We discussed this using the term "psychological tension" in a previous section but now we are getting more scientific.

When we are psychologically uncomfortable because we hold contradictory beliefs, we feel dissonant and experience unpleasant arousal. It is in our human nature, in the very structure of our brains, to strive to decrease this psychological tension. It is an overwhelming desire. Some people are more affected by this than others.

For example, let's say it is drummed into me as a new-hire that safety is my most important priority and always comes first. I notice that other people act in accordance with that charter and I am happy to comply. For three months this is hammered into me, and I fully embody the teachings.

Then there is a month of unexpected bad weather and the work schedule slips and the company will be fined if we don't finish the job on time. I am working in the critical path. Management visits the jobsite more than ever before. I see my supervisor's worried face. Everyone is feeling the pressure. People are upset and everyone steps up their pace wherever they can. I keep hearing about this looming deadline. I'm new. I'm afraid they will replace me with a more experienced worker. I need this job! I still see the "Safety First!" sign every time I enter the jobsite and safety is reinforced at every opportunity. All of a sudden I feel a tension arise, a disharmony within myself.

My subconscious is now working like crazy to minimize my dissonance. The human brain will absolutely decrease this tension. The brain, our subconscious usually diffuses the tension by one of three methods:

- Changes or reduce the importance of our original thoughts.
  - "Safety First!" is changed into "Safety whenever it does not get in the way of getting the job done quickly."

- Give additional strength to an opposing thought (or try and forget one of the thoughts, or rationalize and deny).
  - "I need to take shortcuts if I'm going to keep my job; I need to keep my job. My wife is pregnant. What will she say if I am fired? I need to provide for my family. No one will notice."

  - On and on in an endless loop.

- Let go of an associated behavior.
  - You don't even realize you are not unplugging the saw before you change the blade.

So what do you do as a worker who is committed to safety? How do you fight your own brain? Here is what you do. You must first be aware of the psychological tension you feel on the job and where it is the strongest in the group. Knowing the pressure is there makes you aware that subconscious drivers for accidents may begin to manifest. Start reinforcing the safety messages where the pressures are strong.

We must add one more option that is not explicitly stated in the scientific literature for resolving cognitive dissonance based on our work in the field of innovation:

- Identify the contradiction you and your crews are experiencing and start resolving it. When we resolve contradictions it not only diffuses the tension of cognitive dissonance but creates a feeling of elation. It builds stronger teams and reinforces the priorities.

This now gives you another observational tool: notice contradictions arising. You can be the first to address it strongly. Here is your detailed plan of action to address the team's cognitive dissonance:

- You beat the safety drum even louder and admit to everyone that pressures are rising but safety must be maintained.
  - You talk about it with the worried supervisor and other workers.

  - You make sure that everyone realizes they still have the same goals.
  - If you can do something safe to help with the slipped schedule, you offer to do that.
  - You keep the tension out in the open and not hidden.
  - You constantly reinforce your main priorities.
- You strengthen all of your safety countermeasures.
- You clearly articulate the contradiction and start work to resolve it. You need safety, productivity, on-time performance results, and more.

Studies have shown that not everyone feels the same degree of cognitive dissonance. So frontline supervisors need to learn which types of people will be most sensitive to these forces. *[Remember the importance of getting to know your people.]* Folks with high susceptibility to cognitive dissonance are the folks most likely to do weirdly unsafe things. Once you have learned the specific personality profiles, identify which of your workers fit any of these profiles. Then keep an eye on these workers (keep an eye on everyone, but more on these workers) and help them manage their states. It's been suggested that:

- A person with a high need for certainty and consistency usually feels the effects of cognitive dissonance more than those who are not so bothered by inconsistency.
- Introverts seem to experience more tension and appear to be more likely to change their attitude to match the group point-of-view.
- Extroverts seem to be less affected by the cognitive dissonance and appear to be less likely to change their minds despite majority pressure.
- Self-aware people are less likely to suffer from cognitive dissonance. Watch workers who often justify, rationalize, "second-guess" themselves, look to others for their opinions, or deny past decisions.
- People who are not quite committed or clear on why they are working safely are exhibiting a sign that cognitive dissonance is at work. Committed people are less likely to feel extreme tension.
- New workers who are trying to fit into a workgroup may be more affected.
- People, who hate to be wrong, who deny responsibility, or who limit their intake of new information, are more sensitive to cognitive dissonance.
- People who claim to be victims of circumstance exhibit these typical beliefs and behaviors:
  - Belief in being forced even for simple choices.
  - Belief that it was unfair.
  - Won't accept clear proof of their failures.
  - Claims the victim "asked for it".

  - Shifts blame.
  - Won't accept responsibility.
  - Say things like, "Yes, but... " and "My actions will be justified in the end" even when there is a history of being wrong.
- Highly *sensitive workers* will often display behaviors to over-compensate for sensitivity to cognitive dissonance. You need to keep an eye on them when the priorities on the jobsite become conflicted. They can:
  - Jump to reject alternatives.
  - Avoid or minimize regret.
  - Won't accept clear proof of their failures.
  - Would rather end a relationship than experience a difference of opinion.

It hardly needs to be said, but let's just make sure it is clear. Cognitive dissonance or psychological tension is proof that a contradiction exists. Cognitive dissonance is the result of big contradictions on your jobsite. Tension is the identifier of contradictions. When people do weirdly unsafe things, especially when they know better, it is a sign of cognitive dissonance and it happened because of a safety contradiction. You need to identify the contradiction and relieve the tension that it is causing on the job.

Let's just take this to its inevitable conclusion. Cognitive dissonance is at the heart of safety inconsistency and is the result of the main contradictions in construction. This is a major inventive connection and pinpoints this as a key area to investigate further, as a research topic and for inventive problem-solving.

The relationship between cognitive dissonance and contradictions connects a culture of safety to the need for innovation. It is a very exciting and new opportunity for the crafts worker. If you can learn to identify and resolve contradictions through problem-solving and inventive thinking, you will be very valuable in the changing future of construction.

There are traditional cognitive-behavioral psychotherapy techniques that might be useful onsite. This would also be an excellent topic to explore.

Inaction and Diffusion of Responsibility

There are many situations where people watch a dangerous situation and do not take action to prevent it or to help the injured. None of us like to think that we are that kind of person, but it happens to really decent action-orientated people, too. Why?

It happens so often that scientists have studied it and have named the phenomenon "The Diffusion of Responsibility". In other words, this phenomenon could be called "the watering-down of responsibility" or "the rise of inaction" or the "Why isn't anyone helping me?" effect.

In a nutshell, here it is. The larger the group size, the less likely any particular person will accept responsibility and take helpful action. Add factors such as situational ambiguity, negative social support and strong negative authorities, then people are far less likely to help someone in danger.

How does this apply to safety in construction?

If you have negative leadership and social dynamics against safety, then your people may not act quickly or at all if something unsafe is happening on the job. If the group is large, then any one person will be even less likely to take a stand.

There is one major solution to this challenge which involves awareness of this psychological effect and receiving specific training to counteract its effects.

Part of that solution must involve the mindset of constant, passionate, dedication to safety. We can never let our dedication to safety slip into neutral. We must keep people engaged with their morality, their safety knowledge and define clear intentions on how to be safe. It is so easy to just disengage for a moment and then fall into someone else's dominant mindset: "We have to do this", "We'll get in trouble", "Hurry up", "That will cost too much", "What does Joe think?", "No one else is doing it."

If you have ever stood up for "doing the right thing" with social animosity directed at you, then you understand the difficulty. If you were ever in an unclear situation and other people's actions were ambiguous, you realize why training is necessary. You must have training to understand and practice overcoming the powerful social effect of group size, social negativity and diffusion of responsibility.

We feel that scenario-based training and simulations would be excellent choices to counteract this effect. On the construction site of the future, these types of scenario-based simulation will be common. You will need to interact with this technology and understand the science if you are to be valuable in this future.

We must constantly be ready to take full responsibility and action even if nearby leaders are frozen into inaction, when social forces are ambiguous or you perceive that the group will not support you.

We want our leaders to assess safety situations accurately. We want them to accept full responsibility, take decisive action, and engage others to help when needed. To achieve this, our leaders need to be armed with appropriate training. Whether they can do it or not, you need to train yourself so you will be one of the people who step in to help when the situation demands it.

Leadership is a Strong Countermeasure to the Weird Things People Do

Even if you are the best safety advocate, getting to the next level of performance does require a new level of commitment and leadership. That means change must occur and change can have a different effect on different employees.

Some older or less flexible-minded craft workers may resist strong safety practices because they are already comfortable with less oversight and fewer safety regulations on the job. Sometimes they try to avoid using hard hats or reflective safety vests or 100% tie-off for fall protection because "They didn't have to use them on other projects." These workers with stagnant habits may dismiss the need for proper lock out and tag out procedures around energized equipment or act skeptical about properly sloping a trench or utilizing a trench box as a safety measure, and... so on.

Creating solutions to get total buy-in among generational sub-cultures and diverse opinions is a part of the challenge and opportunity for company management and craft workers. But with gentle pressure applied relentlessly over time, people will accept the protocols as normal and without question; the new generation of the crafts will only know the new way.

When workers are engaged in safety programs it accelerates safety results that last. Great results create success. Positive reinforcement builds on that success and creates strong teams. Everyone loves to be on a winning team, and that creates a positive cycle.

If we can get our co-workers to solve problems together, then the solutions will be effective because solutions sometimes require multiple forms of expertise. Working across departments and operations can expose hidden resistances. Hidden resistance can derail any effort. Great leaders expose the resistance and get their groups to find solutions that are implementable.

Solving safety issues as a team has the additional benefits of strengthening relationships, building ownership of the solution, and reducing resistance.

It is a best practice to face problems early and to include as many stakeholders as possible in the process of solution and adjustment. Remember, people often resist change for legitimate reasons. Each person, department, and operation has their own goals and these competing priorities may create organizational contradictions. We create better results when we listen, understand those reasons and resolve those contradictions through problem-solving rather than putting one team over another.

To lead in safety means to cause the conditions where people are mutually engaged in solving problems and keeping each other safe.

The Power of Observable Behavior in Change Management

Culture is created from observable behaviors and the choices we make in response (mindsets) are self-reinforcing.

Whenever we join a new group, we often hold back. We are watchful. Why is this? We intuitively know that to join a group we must understand the important factors of interaction so that our behaviors and thinking are not wildly disconnected from group norms. If we behave too differently, we may be rejected, ostracized, or judged to be lacking and that can lead to having low credibility with the group. We would not be relevant, and therefore, we would not be valued. So, we observe the leaders and most influential people within the group. We gauge the range of behaviors that are acceptable. We gauge the diversity of language that others use. We model the behaviors we see from our leaders and we get a good start in the group. We are accepted and no longer have to hold back. Now, our opinion matters.

We also inadvertently begin joining a group mindset in which we reinforce each others' thoughts and behaviors. This makes change difficult to implement. Sometimes employees will refer to the company culture as "they" ("If 'they' think I am going to do this..."). Sub-textual messages are powerful. There is hidden power in the CHOICE TO CONFORM to a group. We are strongly influenced by our observations of behavior and our belief in the consequences of the failure to conform.

Strangely, this simple process occurs in reverse, too. If an outsider joins a group and does not conform but is highly effective and yet somehow also socially fits, the organizational leaders will be very responsive to the newcomer. In some cases, the culture will strike back if the person is effective but breaks too many sacred social norms.

Effectiveness in leading change requires perceived (real is better) congruence with beliefs, mindset and action. In other words, you must believe as you act, and you talk as your integrity demands. All of you should be in accord *(Remember what happens when parts of you are in tension with other beliefs you hold?)*. This acts as a radio transmitter that delivers an authentic message. It may present and percolate in a subliminal way, but everyone receives the message. This may be a challenge for leaders and may involve substantial personal adaptation in style, mindset, belief or action, but the results are dramatically useful. When people say, "Walk your talk," this is what they mean.

However, the power of observable behavior can have a dark consequence, too. Once an observed leader compromises on something important (like safety) - regardless of the reason - then like a radio transmitter that delivers your message of hypocrisy, your observed message will percolate in a subliminal way, but everyone receives the message.

The Perfect Storm for Change

Our experiences have shown us that the perfect storm for change occurs when:

- A compelling and pressing reason for change is imminent. Facing risk causes people to change their approach.
- A new mindset and approach is formulated that promises a better operating perception of reality that can address the new problems and challenges being faced.
- High-profile, influential leaders commit to make the desired change happen.
- These leaders actively model the new mindset and behaviors and produce new results, which demonstrate the potential opportunity being realized. In other words, they lead by example, which is very compelling!

Because culture is formed by observing the behavior of leaders, others will make better choices as they notice the new way. It is good for this to occur naturally so the culture has time to adjust. Thoroughness creates confidence whereas speed with missteps can destroy confidence. Change that occurs through force, fads, or flawed programs often leaves scars and even damages the credibility of current and future leaders.

Some people passionately hate change and even good change will be painful for them because they have difficulty adopting to new methods and behaviors, or producing new results. They fear losing their job. Friends and co-workers may feel their pain and react negatively to support them.

We believe that in order to gain the cooperation of a group when facing imminent or required change, the process must have a shared sense of responsibility, vulnerability, and accountability. If you want people to feel assured that leaders mean business and that change is really going to happen, your team will need to observe the following:

- Admirable leadership,
- Transparency and the truth,
- Authenticity within the process,
- Fairness and connectedness,
- Communication and receptivity to feedback,
- Goals that are achievable and that address the risks,
- Shared vulnerability and accountability for all,
- Problem-solving and positive results.

So, change is a process of walking through uncertainty and discomfort. However, if the new way is very compelling, and the leader's ability to model and deliver results is strong, change will occur faster and with far less resistance. Modeling leadership for the key stakeholders and social network influencers helps the entire group achieve change faster. As you can imagine, this is vital for safety in all work environments.

Using the Pace of Change and Resistance

Changes in group dynamics usually occurs very slowly at first and then progresses more rapidly.

- It happens one person at a time, starting with a leader (whether they have delegated authority or not).
- It occurs in a person's heart as a decision. Therefore, understanding how the change affects you as an individual allows the choice to occur.
- The choice to embrace overall change may not encompass the full range of desired behaviors and mindsets.
- When about a 1/3 of the group makes the shift, the rest adopt it more rapidly or choose to leave.
- The pace of change accelerates when a sense of urgency is embedded into the culture.
- Some will fight for the established or "old" way. They usually (about 95% of the time) offer insights about what is valuable and should not change. Do not ride roughshod over this resource. Do not change good parts of the culture; incorporate the new synergistically with the old.

For leaders, the process is different because they know change must happen and want to make it happen, but they also face important issues raised by the community. When a leader starts the change process and resistance crops up, there can be a tendency to become controlling, fearful, or try to avoid criticism. When issues get tough there can be a tendency to separate from the issue and to simply focus on insisting on change. In dogmatic organizations, this approach may partially work because people are used to it. But to have seamless change, a leader must accept that they are as vulnerable in the change process as are their co-workers. They must model this, acknowledge the difficulties and use problem-solving tools to resolve issues for good. This means that the best way to cause change to happen is in an acknowledged vulnerable state while armed with tools to counteract the vulnerability. What a beautiful paradox. This is mutual engagement versus an antagonistic relationship.

Construction people are "rough and tough", so you are probably thinking, "What?" and "No way; that sounds weak." Here is a more robust (and just as truthful) way of looking at it. Our approach is to treat change like a risky journey. You will need good tools to help you along the way. You must be able to predict the risks you may face.

Retire or eliminate your risks as early as possible. Be agile in the face of unknown danger. Use all of your resources carefully. Be thorough and complete your mission, because backtracking is expensive and can be dangerous. Do it together as a team and rely on each other.

Since identifying all hidden problems and resistances is important to success, you need a basic problem-solving toolset (innovation skills) to support your team and organization.

The Role of Enthusiastic Evangelism – You Change First and Solve Problems

We love our Structured Innovation toolset and we are passionate about safety so you may be surprised to read that we shun strong fan-fare. We have found that people at all levels can be very negative about "junk" or "fads" being fed to them by people unwilling to get in the trenches and solve real problems. We have never found a situation where it was better to force the people to just "drink the Kool-Aid". New best practices *must be adapted* to the existing culture if you want them to be effective.

It creates unnecessary resistance within your crews if you try to sell them a program. Don't sell - INFORM. Involve key influencers in the acceptance process and prove the value of problem-solving step-by-step. These people will become almost fanatical for a true and solid solution and they will pass it on naturally. When anyone "walks the talk", it is much more effective than pressure.

People don't like to be SOLD, they want proof - want to see the results - and they want safe passage to the new way. It takes time to create buy-in, just like it takes time to develop credibility in a relationship.

- Do not push your culture and ask people to believe blindly.
- Look for the best place to achieve great results.
- Find your early adopters.
- Keep an open-mind and foster a spirit of inquiry.
- The tools we have given you are pliable. You can fit them into your organization. It is unwise to try and fit your organization to the tools.

After the entire organization has embraced the change and put their stamp of approval on it and are delivering solid results that cannot be denied, then consider some banner waving.

Priorities

Lots of companies and people talk about what is important to them. That is on the surface.

What is prioritized and what is not prioritized reveals important sub-textual information. This is deeper down. This is not usually talked about. This is more obvious to others than it is to ourselves under self-examination. This is more of a self definition. This is where we will actually act from. This is a lot more important than what we say. To be a person that others respect, the priorities we act from should be the same as what we say. Put simply, we must walk our talk.

What we say, what we do, what we choose, what we admire and reward, what we research and learn, and what we value must be aligned and supporting the other. When we are at war within ourselves (what we say is different than what we want or believe or do) then we will put out some serious mixed messages and we will do some strange things.

It is essential that we set clear expectations. It is just as clear that we can never tolerate compromise in our priorities. In this, our head, heart, and gut must be in alignment.

Final Thoughts on a Culture of Safety

Since observable behavior is the cue that we use to discern group norms, we think that it would be highly advantageous for you to keep track of this topic on the web and blog about it.

Also, keep track of your own observable group behaviors. Realize you need to maintain a positive group psychology for your culture of safety that is always moving toward safety mastery. Strongly communicate that culture. It begins with positive and fully-committed leadership.

Change only the most critical aspects of culture that block acceptance of the new initiatives. Typically less than 5% of the behaviors in a culture actually interfere with a culture change[64]. Understand your organization's capacity to change by evaluating its readiness for change.

---

[64] From a conversation about culture change (about 1998) in Sacramento, CA with Price Pritchett, culture change thought leader and mergers and acquisitions consultant.

Scott Burr, a co-author of this book, has used this principle with success to improve acceptance of Six Sigma and Lean implementations.

Robert G. Eccles, Ioannis Ioannou, George Serafeim; "*Working Paper - The Impact of Corporate Sustainability on Organizational Processes and Performance*" July2013 http://www.hbs.edu/faculty/Publication%20Files/12-035_a3c1f5d8-452d-4b48-9a49-812424424cc2.pdf (accessed 31august2013)

People do weird things because they experience contradictions as psychological tension. The scientific term for psychological tension is 'cognitive dissonance'.

Identify contradictions created by change and resolve those contradictions using inputs from any dissenting groups. Encourage others to speak up, listen to the input. This gives you enormous power and value to the group. This honors the real reasons behind resistance to change. But help define that resistance so it is precisely negative and therefore it is useful. Use a team-based process to solve problems that involve multiple disciplines and create breakthrough innovations for change. This is a unifying factor in culture.

Every organization is a complex system. Understanding the dynamics of that system supports the change process. It only seems easier to bully change down the ranks - but it's not! Bullying takes constant concerted effort. If you want to prevent a well-intended change from creating the opposite of what you need, understand the hidden mindsets and resistances in your culture and address them.

All organizations evolve and change. Understanding the basic contradictions of your organization creates a point of stability, focus, and action that can help you make real progress over time and aligns all of your actions with your corporate safety goals.

Of course it all occurs through leadership which is a powerful countermeasure to keep safety and other valuable business objectives moving forward.

By applying these basics, you can achieve dramatic positive change and be rewarded with swift results. The lives of your co-workers depend on it.

Bottom line #16: ***Stress, fear, uncertainty and awkward social situations drive unusual automatic behaviors and create fuzzy thinking in people. If we are to achieve 100% zero harm we must understand and resolve these underlying problems.***

Safety Mantra #16: ***People do weird things when they are afraid and unprepared.***

Take-Away Points: ***Chapter 16***

- Train yourself to recognize and handle the strange behaviors of people: normalization of deviance (lying to yourself), foolish obedience, fear of being wrong or criticized, situational uncertainty, unclear social situations, negative social support and mindsets, knowing better but doing it anyway, and inaction and diffusion of responsibility.
- The ultimate countermeasure to addressing stress, fear, uncertainty and unclear social situations is awareness, training and positive leadership.

- Solving safety issues related to stress, fear, uncertainty, and unclear social situations as a team is the goal.
- Team-based solutions have the additional benefits of strengthening relationships, building ownership of the solution, reducing resistance, and removing hidden messages and replacing them with observable ones.
- Culture is created from observable behaviors. Every positive choice we make in response to a positive observation is self-reinforcing.
- You must display admirable leadership, because you will be observed by others: be truthful, authentic, fair, connected, receptive to feedback, accountable, and able to recognize and solve problems.
- Remember that situations change and organizations evolve. You will need to constantly pay attention if you want to align your actions with your safety goals.
- You and every other craft worker drive safety change with a sense of urgency and honor resistance. Listen to your culture's precise negativity. Find the solutions buried in it and take action on all you know.

"Many of life's failures are people who did not realize how close they were to success when they gave up."

- Thomas Edison

"There's no stopping a man who knows he's in the right and keeps a-comin'."

- Louis L'Amour, Spoken by Flagaen Sackett in Galloway

Chapter 17 – Safety Leader Behaviors, Mindsets and Character

Much of this book has been about taking responsibility for what you choose to do, whom you choose to be, and for the mindset you choose to display. Taking responsibility is a hallmark of great leaders and great problem-solvers. In truth, we are asking you to become both a great leader and an excellent problem-solver in order to make your work environments 100% safe.

Mindsets of Great Safety Leaders

Great safety leaders seem to embody a passion for safety beyond compare. They stand for safety first, act for safety always and refuse to compromise. Zero harm is the goal and they are not satisfied with anything less than that. They also have a low tolerance for excuses.

We are now asking safety leaders to add other business values to the safety equation and be the central point where the toughest organizational problems are solved. You are charged with maintaining a no-compromise mindset, great problem-solving skills and a bias towards safety while advancing productivity. Safety is the center point where you experience contradictions, therefore it is the best place to find the highest payoff problems and generate the most effective solutions.

The quality professionals in manufacturing took on this task, succeeded grandly, and did it without compromise. Now it is your turn.

Great safety leaders are purpose-driven, persistent, dedicated and absolutely certain that the goal (zero harm) is achievable and *must* be achieved.

They know that the courage to face problems and stand up for what is right saves lives!

Behaviors of Great Safety Leaders:

- Exhibits caring power.

- Gives respect. Because they are congruent, they earn respect, too.
- Encourages people to speak up.
- Is a good listener.
- Gets to know people.
- Is open to safety suggestions.
- Is always a morale builder.
- Praises publicly and scolds privately.
- Realizes that people are the greatest asset.
- Always leads by example.
- Strongly promotes workplace safety buy-in.
- Good communicator of company policies.
- Publicizes lessons learned.
- Recognizes and promotes other safety leaders.
- Gets to the root cause of an accident or incident.
- Takes action if an unsafe act or condition is observed.
- Coaches and trains on safety.
- Mentors at every opportunity.

Characteristics of Great Safety Leaders:

- Confident and capable in one's capacity.
- Holds oneself and others accountable.
- Forward thinker and problem-solver.
- Unwavering loyalty to first principles of safety.
- Goes the extra safety mile.
- Values and defends safety integrity.
- Has an upbeat and positive outlook.
- Is open and honest.
- Is confident and consistent.
- Is compassionate and dependable.
- Is fearless in pursuit of workplace safety.
- Keeps driving towards safety perfection.

Traits of Great Problem-Solvers (Contradiction Solvers)

Leaders and problem-solvers have many things in common. Among the most important traits is the natural assumption of responsibility. Both leaders and problem-solvers monitor the mindset, goals, and signals that point to obstacles. They also accept responsibility for challenging assumptions that create the As-Is situation. And they are responsible for believing a more ideal situation is possible. A sense of responsibility is one of the most important traits you need to develop in order to achieve your safety goals.

The underpinning of most construction companies and industry regulatory organizations is reactivity. However, proactive problem-solving is the key to success and that takes a specific set of personal traits.

We start developing problem-solving skills in ourselves and as a core competency within our companies by promoting consistent mindsets, characteristics and behaviors that enable highly-effective leaders and problem-solvers.

So what mindsets, characteristics, and behaviors are most valuable for solving problems? Effective, serial problem-solvers:

- Understand the power of a useful mindset.
- Take responsibility for the current situation.
- Pursue the most ideal solution.
- Have the courage, discipline, and will to proceed.
- Have problem-solving persistence.
- Seek advanced resource productivity.
- Understand the power of analogous thinking.
- Can switch between a system view and detailed view.
- Challenges beliefs and facts.
- Solicits input from all stakeholders.
- Leverages the expertise of individuals and teams.

A mindset is something to be used to your advantage. Every mindset has an associated advantage. You must know what it is you want to create in order to make use of the correct mindset.

Of course, good problem-solvers are competent at their job and readily learn from successes and failures, which they are happy to share. They also learn from the past and from analogous experiences. They are able to keep track of data and validate facts yet are objective and flexible. They think systematically, see patterns and connections, and keep the focus on important requirements. They are able to set priorities and manage risk.

Some of the important attitudes and attributes include a sense of self-confidence and an awareness of personal limitations as well as the ability to gain trust, respect, and credibility. The great solution-makers are continuously curious, adaptable, realistic, and positive as well as dedicated to ideality's success. They are adept at the art of questioning and seek to expand their information base, advance ideas, and create a more comprehensive view of reality. Great problem-solvers are effective even when uncertain, are intuitive, and are skilled at their job, but also able to see beyond their boundaries.

Another important set of skills involve leadership. In order to solve the last 5% of entrenched safety problems, a person must value and recognize others, build team cohesion, improve the dynamics of their team and see their situation objectively. This can be accomplished by leaders at any level of authority. They must be able to communicate the direction they wish to go, influence their peers, and enroll others in their vision. When needed, they are able to coach or mentor others, feel comfortable and yet are able to delegate to others, and make good use of available resources.

Great safety experts and problem-solvers are proficient communicators. They value personal interactions, use many modes of communication (visual, experiential, verbal, kinesthetic, analogy and storytelling, etc.) They listen effectively and are able to translate the information into useful messages and relay it forward. They value open and honest communication. And finally, they are comfortable making decisions that do not compromise safety.

It's interesting that some of the most important traits in really talented problem solvers are integrated contradictions. For example, great problem-solvers have the appearance of contradictory traits. They are both:

- Humble and fiercely resolved,[65]
- Flexible and rigid,
- Fearlessly cautious,
- Faithful and questioning,

[65] Source: We are presenting this information through the lens of the language created by Jim Collins, a well known and respected business thought leader. The source of this information is Innovation Scientific's application of TRIZ knowledge in a novel way: Use the lens of contradictions to manage and advance the competitive position of individuals and businesses.

- Creative yet disciplined,
- Simultaneously put people and the plan first,
- Achieve mastery on the cutting-edge and stay focused on routine challenges,
- Have monopolistic knowledge and strategically share power.

But these traits are not contradictory in the heart and mind of the innovator. They are all traits that are held in the tensionless place - I WILL SOLVE THIS PROBLEM!

Great problem-solvers simultaneously hold beliefs that appear to be contradictory:

- In the current situation there is no known solution and the situation is bleak and dismal... AND
- In our future reality, we will have a solution; I may not know how to do it yet but I will continue until I prevail.

In essence, the innovator embraces cognitive dissonance as a way to deliver relentless pressure and psychological tension so that they can marshal the resolve to address the problem.

People who are *not* great problem-solvers usually hold very different beliefs about their tough problems. Some of those false beliefs about problem are:

- It's not worth solving.
- I'm not doing that.
- It's not really a problem; it's just the way it is.
- It can't be done.
- It's a product of the organization and the result of previous decisions.
- The opposition is unmovable.
- There is no opportunity; it is a problem.

The authors of this book believe the current mindset of most construction industry leaders is to hold these false beliefs in stubborn disregard of the truth.

Many people fall into this category and are stopped easily by the slightest obstacle. You must flex your persistence muscles in order to have the will to move forward. The authors believe that problem solving courage is the true tradition of the construction industry.

Problem-Solving Courage and Persistence

Two foundational competencies great problem-solvers consistently exhibit are the *courage* to begin and the *persistence (tenacity)* to keep going until success is achieved. great problem-solvers take responsibility to work on high-value problems that seem to be 'too overwhelming' to others. Their solutions will redefine industry beliefs.

When there is no map, no mentor, no understanding of how to proceed, it takes courage and determined commitment to enter into the chaos of multiple possibilities and unknown pitfalls. Rules do not yet exist and choices have not yet been made, thus there are few limitations.

"Next step" or "incremental" improvements do not demand the same skill set or rigor as doing something no one else thinks is possible. Many competent people are comfortable only when a plan is laid out. They feel persistent only if the territory is familiar, or if there is a mentor they can question and whom they trust, or if the procedural steps are visible and laid out for scrutiny - if the way is logical. As long as the way is clearly defined, traditional thinkers are willing to proceed. The job appears exponentially more difficult when the way is not clear.

Just like any discipline, the more you proceed and succeed, the more you build trust and belief in yourself. Then it is easier to begin the next time. Like a journey, it begins with the first step.

Great problem-solvers and leaders do not let fears stop them - and that is neither unrealistic nor unduly optimistic. They embrace a unique paradox: 1) face a great problem without knowing the outcome and 2) possess absolute resolve to overcome that obstacle. [66] In order to create a breakthrough, you must have the persistence and the resolve to proceed while remaining practical and grounded in the current real situation.

> *"There is no chance, no destiny, no fate that can circumvent or hinder or control the firm resolve of a determined soul." - Ella Wheeler Wilcox*

Responsibility

We could try to solve a problem using a traditional methodology based on historical

---

[66] We are using the language of Jim Collins. In TRIZ it is called pursuing ideality. Innovation Scientific defined the term "ideality's essential tension", which defines the contradiction between challenging the as-is and achieving the to-be condition. Most people appreciate Jim Collin's clear and simple way of expressing these concepts and so we have applied his leadership language to great problem-solvers.

experience, well-honed skills, and established mindsets. Or we could use an innovative approach.

The conventional action plan might be as follows:

- "If it ain't broke, don't fix it"
- When it breaks, figure out who or what is to blame;
- Prioritize which parts need to be fixed;
- Draw upon experience or look around for who can help; "What's new?"
- Figure out "Who is standing in my way?" or "Who needs to get on-board?"
- And calculate how much of the solution we can afford to implement; "You can't have it all."
- Then take logical, obvious actions.

As we can see, the subtext of an orthodox action plan is one of *reactivity* and a *trade-off* mentality. This can drive traditional problem-solvers into corners they can't think their way out of.

From the onset, a compromise-based subtext is very different from a problem-solving subtext, which is *responsibility* and *persistence*. See the difference:

- Determine "What's not working now?" or "What's going to stop working?" or "Where can we improve?"
- Determine what we want by asking, "How can I get more of what I want AND less of what I don't want?"
- Decide whether it would be better to draw upon experience, look around for what's new, or take a structured approach to create a new more beneficial outcome.
- Interact and facilitate with key stakeholders to create a robust solution (this is a natural approach to change management). Together with your team of key people address secondary problems and predict and resolve potential failure before it happens.
- Then implement using logical, obvious actions.

From the onset, it is the problem-solver's responsibility to keep tabs on (manage) the mindset and monitor goals and recognize signals that point to obstacles. The innovator is also responsible for challenging the assumptions that created the current situation. And the problem-solver is also responsible for believing a more ideal situation is possible. Be responsible for your solutions.

Don't try to solve a problem; don't hope; don't think a better solution may be possible. Know or at least act as if the solution exists and retire uncertainty one step at a time. And, all the while, keep tabs on and manage the current mindset.

Great problem-solvers manage the "mindset inertia" as both a team and personal asset. These leaders identify which important problems to solve, utilize the tension between resistance and opportunity, remove false assumptions and beliefs, assess risks, and use a structured methodology as a science.

Great problem-solvers:

- Pursue ideal solutions - Push "Ideality's Essential Tension", which is the tension between pushing for the ideal solution and getting it (or failing because it was out of reach).
- Resist the urge to compromise or trade-off or give up.
- Persist when faced with resistance.
- Find uncompromising solutions to competing forces within a problem (resolve contradictions).

Great problem-solvers are aware of system resources and know how to find resources hidden within the problem. They constantly look for resource productivity within their solutions. They achieve the same functionality while eliminating problems and use only the resources locked within the system. Another way to achieve resource productivity is to reduce the scale of resource usage (small is often more efficient). Resourceful solutions have multiple benefits (rather than trade-offs) such as greater safety plus enhanced corporate image, improved strategy, and lower cost.

Great problem-solvers take into account the organizational, political, or social constraints within the problem as necessary. They do not fight organizational resistance; they use it to strengthen the solutions, but they also know that acceptance and commitment at all levels of the organization is eventually vital to success and solutions sought. They accept responsibility for all aspects of the problem, even if those aspects are "outside" of their sphere of influence.

Summary

Great problem-solving leaders have a resolute expectation of producing positive results in the highest sense and take full responsibility for the current situation no matter how dismal it is.

In our modern world, people who embody these mindsets, beliefs and behaviors are precious and should be valued as indispensable by companies that are serious about safety perfection. Those who want to stand out realize situations change and safety is always under construction.

Bottom line #17: ***Safety perfection begins with a problem-solving leader who accepts responsibility, resolves contradictions, and persists until both a 100% zero harm environment and productive processes are delivered.*** **This leader must have a** ***process focus and a systems viewpoint. Learning is now a part of the new safety and construction mindset because continuous safety improvement is mandatory for survival.***

Safety Mantra #17: ***Safety perfection is a journey of safety mastery.***

Take-Away Points: ***Chapter 17***

- Great safety leaders embody a passion, act safety, refuse to compromise, and are more preventative, scientific, and process oriented than their peers.
- Great safety leaders are purpose-driven, courageous, persistent, dedicated and absolutely certain that the goal (zero harm) is achievable and *must* be achieved.
- Safety requires leaders to accept responsibility, exhibit caring power, and build team cohesion.
- Develop both humility and fierce resolve.
- Be fearlessly cautious.
- The new focal point for leadership in construction is the safety organization.
- Be disciplined and adaptable. All these "contradictory" traits are held in a tensionless place - I WILL SOLVE THIS PROBLEM!

PART IV – A CALL TO ACTION

The best way to predict the future is to invent it.

- Alan Kay

"If you have built castles in the air, your work need not be lost; that is where they should be. Now put the foundations under them."

- Henry David Thoreau, Walden

Although there are bright spots in the industry today, the prevailing condition within the construction industry is bleak because people continue to be injured or killed on the job. The attitudes and hidden mindsets that manifest this condition are:

- Safety is seen to slow down and even interfere with production. *You now know this does not have to be true.*
- People believe that everything that can be done is being done. *You now know this is a delusion.*
- Workers 'dodge bullets' and think that is normal. *You know that this leads to accidents and embeds this deviance like a virus into the mindset of future construction workers.*
- People behave strangely; don't think ahead; don't always do the right thing. *Now you know why and can take steps to stop yourself from doing it.*

The fact of the matter is you *can* get to zero harm if you approach the goal of safety perfection scientifically, systematically, using problem-solving skills, resolving contradictions, and persisting until you get what you want.

Because you read this book, you will have better tools, behaviors, and mindsets to help solve the most confounded issues on the worksite. And that will keep you relevant as the industry changes.

We expect that you now think differently about accidents, safety, and risk. You realize more safety gains are possible. For example, if you hear a worker say, "I don't know why I did that; it was stupid and I know better; oh well, accidents happen." What will you think? Or, even more importantly, what will you say to that person? We bet you give a different answer today than you would have given before reading this book.

One more thing; it's time to prepare yourself.

"Up, sluggard, and waste not life; in the grave will be sleeping enough."

- Benjamin Franklin

"The price of greatness is responsibility."

- Winston Churchill

Chapter 18 – How to Prepare Yourself

So now you know the construction industry is going to become more technology-driven and science-based. The industry will develop more standards than ever before and require workers to be more engaged with them. Safety will be built into construction systems and these systems will be pervasive, predictive, interact in real-time, and they will collect and analyze data from project conception through decommission. In fact, useful and precise data will be a competitive advantage in the marketplace. All projects will be process-focused. Contradiction management will be a key job description.

You may not yet notice any indications of these changes but they are coming and some of the changes will put construction workers out of a job. If you want to be considered a valuable asset in the future, you will need to modify you current practices. You will need to accept responsibility for more aspects of safety and production and you will need to think differently about how you are adding value to your employers.

The authors' ideal goal for construction businesses is: *"Processes and projects will have safety and quality built-in, with cost and performance as prime considerations and meaningful work for all."*

One More Shot at Explaining the Value of Contradictions

There is a lot of excellent value in this book, but the concept of recognizing and resolving contradictions may be the most important single idea in this book.

The ability to resolve contradictions is so universally needed (across so many disciplines within construction and business) you can use this single mindset and skill to be valuable anywhere in life.

So please know this and live it!

Contradictions, compromises, optimizations, and trade-offs are opportunities. The bigger and more fundamental the problem is, the more valuable the solution will be. Resolving companywide and industry-scale contradictions will make you a precious and valuable contributor.

Technology Will Be Everywhere

You will need to interact with many types of technology. Get comfortable with learning new technology-based software, hardware, tools, ideas, and information.

Get used to being "watched". Data streams are going to collect information about everything on the worksite, including you. The data is going to be used systematically to ensure maximum use of resources, maximum safety, maximum quality, and profits.

Information about your job, the conditions, behaviors, and risks will be simulated long before you pick up a tool - think of it as a JHA on steroids. Communication will be pervasive, too. Even the clothes you wear will be activated to save you from a fall or notify you of a possible danger. But even though the system is picking up the responsibility for safety, you will be required to be more responsible than ever before and interact with process controls for every part of your job. The system cannot become the 'cop'.

Learn More, Learn Continually, Grow

Start an education program for yourself. Learn new construction-related tools and techniques, of course; but also learn new computer-based tools or applied sciences, or social-based research. We've suggested many throughout the book, but find what areas are interesting to you and think about how those would add value to your work and to safety.

You will understand the difference between countermeasures and solutions and be responsible to strengthen safety programs on a continual basis.

You will need to improve your decision-making skills.

Attitude

You must learn to expose and change your hidden patterns and unproductive mindsets. You can also accept responsibility to support the development and implement of a mistake-proofing[67] system, which will require a cultural change.

---

[67] Mistake -proofing is a strategy to make processes mistake proof. This was developed by the Toyota Production System.

You must know the "hidden" triggers of accidents. You must control your "automatic responses" and manage your mind, your attitudes, and your tensions. No longer will people be allowed to do weird things and have "accidents".

You need to develop the motivation to 'own' problems. You need to believe that you are more than a pair of hands in a reflective vest; you have a more meaningful contribution to make. You observe, understand, and take an active role on the job. You help others on the team and have a vision for your success. Resolving contradictions is your responsibility. Uncompromisingly you take that responsibility to heart and commit to being a solution-seeking person.

You will not participate in group stink think.

You will not blame or finger-point or conceal mistakes.

You must learn to embrace positive change. If the requirements of your job change, embrace the change rather than resist it. If the change is going to be detrimental (like automation that gets rid of jobs) you had better have seen it coming and worked with your community (new relationships and old relationships used in new ways) to influence the changes. If you saw nothing, knew nothing, and did nothing, then you have no right to bemoan the results that you helped create by your inaction.

Use the Quality Movement as an Analogy

What does a middle-class worker want? S/he wants a steady job that allows the bread winner to come home safely every day and provide the family a house in a safe environment. They want to put the kids through school and create enough savings so they can live to the end of life in a relatively secure fashion without unduly depending on others. The hallmark traits of the American middle-class are hardworking, responsible, and somewhat independent. They are proud to earn their living.

Seemingly at odds with that are the business tools - Six Sigma, Lean, etc. - that are very good at increasing productivity and eliminating jobs. In fact, practitioners are so good at this, it seems like it is natural and inevitable. Of course, this often means that workers are down-sized. Many leaders believe this trade-off is inevitable. We now know that it does not have to be.

However, if productivity gains are always achieved at the cost of jobs, then eventually a negative tipping point will be reached and that will hurt the common good. Manufacturing jobs plummeted as a result of productivity improvements. The gap between the rich and the poor grew. Some argue this is still happening.

The idea here is that the common good must be guarded. It must be watched, discussed and brought out of the shadows. It should be of major interest to many diverse groups with different objectives.

NOTE: We expect a similar level of drama in construction as was experienced in the manufacturing quality movement. There will be winners and losers and the common good may or may not be served. Now that you have this analogy you can benefit from it. You will need a worker's-based strategy to benefit from the inevitable changes coming. If the traditional outlook of business prevails, as it did in the quality revolution, then construction jobs will be eliminated through process improvements.

Therefore, it is the responsibility of every member of a community to watch the reality of your situation and to be ready to innovate, solve contradictions and clearly examine the psychological inertia within the stories all sides are telling.

Each person capable of leading for the benefit of the community is a valuable resource.

It is up to each person to remain relevant: for themselves, their family, and their community, for their employer (or employees), and for their government/nation. You need to develop a high-fidelity understanding of the situation. You must base your beliefs on solid research and not get caught up in some of the dysfunctional social patterns we explored in this book. Make sure your grasp on reality is not being fed to you by people who benefit from over-stimulating you rather than educating you.

Start with examining *your* mindset inertia and developing *your* contradiction-solving skills.

The quality movement made companies more productive. They also cost a lot of American workers their jobs. If you do not want this to happen again in your industry when the safety movement begins, you need to solve some of the current problems differently than the quality movement solved their similar problems. Learn from the worker's (and union's) failures.

You can also learn from some of the successes.

One thing the quality movement did very well was to create a *call to action.* Worker organizations can learn from 70+ years of this call to action. Model what worked then, understand why it worked, and figure out how that is applicable today in the current situation. Construction workers must make their Safety Call to Action very effective AND inclusive of workers. Several reasonable avenues exist to achieve a 100% zero harm record with worker involvement and increased business productivity.

So, how can the worker be more proactive? Involve a wide community in the safety call to action. Get help!

Engaging with Other Leaders

Other groups that speak for the whole society (other than regulatory agencies) belong to some very special groups:

- Non-profits,
- Media (including blogs and internet venues),
- Watchdog organizations,
- Engineering societies and science organizations.

These organizations played a key role in the evolution of quality in manufacturing and thus can do the same for safety in construction. Get involved. Get them involved with your issues. Do it now even when change is way off on the horizon. We challenge workers to engage with these community leaders to take the safety conversation to a new level.

The leaders of these groups, when they are performing the highest order of their responsibility, hold the interests of the whole. It is a challenging job because many of the priorities of specific groups in society are contradictory and that can create friction and conflict. But, since you are developing your skill on resolving friction and conflict, you can help these groups work synergistically for the worker's agenda.

These intermediary groups open conversations, bring perspective, and keep the proper focus for the whole: this means vigorous high-spirited discussion on the value of safety perfection, business productivity, and worker jobs! We want people arguing all aspects of this issue. We want people to disagree and bring up contradictory points-of-view. All of that arguing and debate provides valuable input when it comes time to resolve the contradictions. Then we can solve the problems.

Examples of some of these groups and include:

- American Society of Safety Engineers
- American Society of Mechanical Engineers
- American Society of Civil Engineers
- National Association of Safety Professionals
- International Safety Equipment Association
- Rocky Mountain Institute
- Associated General Contractors of America
- Bill Moyers and Company
- Pew Research Center
- BBC, NPR, and PBS
- MacArthur Foundation

These people and organizations have the passion to express intolerance with current results. Demand leadership. Ask why safety mastery is not a priority along with American productivity and jobs here at home.

One other resource is a professional innovator. Reading about Structured Innovation could help you and your other groups understand what is happening using different more productive points-of-view. Professionals using advanced Structured Innovation could really make a difference with crafting an agenda for safety while simultaneously enhancing productivity and job retention. By managing market contradictions and applying advanced tools, such as evolutionary patterns and the technological lines of evolution[68], worker organizations would be able to predict clashes between the priorities of their stakeholders. They could establish strategic directions long before the industry realized there was a need and could create value from this by:

- Delivering value on the right priorities,
- Improving the use of resources,
- Aligning forces where possible; creating important "partnerships" and synergies,
- Anticipating and surviving changing political forces,
- Utilizing public, political, and scientific-community criticisms to better understand and align with their mission,
- Developing communication strategies to pitch their value to the public before, during and after a crisis.

Summary of Advice

Here is the summary of our advice to you so far:

- Safety perfection is the goal - every worker goes home safe every day.
- Quit thinking in compromises. The safety versus production trade-off is dead. Those who quit indulging in this compromise will succeed and be valuable. Those who don't belong to the walking dead.
- Learn to identify, solve, and manage contradictions, compromises, and trade-offs.
- Be more process-oriented and learn planning skills.
- Think before acting. Recognize when you have a 'trigger' response.
- Deal with the weird things that you do through awareness, training, and leadership.

---

[68] The evolutionary patterns and lines of evolution are just a few topics that can enhance an understanding of where to evolve an organization. These are sub-topics of Directed Evolution (TRIZ) or Competitive Opportunity Management (Structured Innovation) toolsets.

- Think in terms of process, solutions, and inventive ideas to address contradictions.
- There will be more reporting, more data collection, more recordkeeping, and statistical analysis. Be open to that. Participate with it.
- Some jobs will require more discipline; some jobs will require more creativity. To be very valuable in the future, you will need to have both.
- Your company will deploy new processes and new initiatives. To be relevant you will need to be flexible, responsible, and constantly add value. This is the way to keep your job. Be at the cutting edge if you want to be valuable in the future.
- Join with other leaders and organizations to identify and resolve the biggest safety contradictions in your company and industry.

The Goal: Safety Perfection and the Journey of Safety Mastery

As the authors, we truly believe that our body, mind, and heart contributions can make a dramatic difference regarding safety in the workplace. Assimilation of knowledge gained from this book, coupled with right action, will result in untold numbers of people avoiding the ultimate tragedy - being killed or injured on the job. We may never be able to measure what does not happen - and what a great thing that truly is - because many "accidents" just won't happen. Think about that.

Now expand your thinking across industries and worldwide; envision mega-millions of "accidents" that will never happen.

We ask that you join us by investing your personal energy and commitment to get this book out to everyone within your sphere of influence as a part of your effort for achieving safety perfection! Talk about the issues raised in this book. Post blogs about the importance of solving problems. Write articles. Discuss the character it takes to stand firm for safety and the many countermeasures discussed in this book. Please do cite us as the source.

May your courage, character, and heart help you achieve outstanding safety success. We hope you have the will to persist as this book contributes to the way.

Engage, participate, and partner in building upon this paradigm shift for the sake of the people within your organization and to wherever your sphere of influence extends. Safety leadership at any level of authority coupled with persistence and inventive power aligned with worker buy-in, can become a powerful dynamic that is rewarded with dramatic safety success, improved productivity, and meaningful work for all. It can be rapid. It can be real, and it never has to be rare again.

"It is not the critic who counts: not the man who points out how the strong man stumbles, or where the doer of deeds could have done them better. The credit belongs to the man who is actually in the arena, whose face is marred by dust and sweat and blood, who strives valiantly; who errs and comes short again and again; because there is not effort without error and shortcomings; but who does actually strive to do the deed; who knows the great enthusiasm, the great devotion, who spends himself in a worthy cause, who at the best knows in the end the triumph of high achievement and who at the worst, if he fails, at least he fails while daring greatly, so that his place shall never be with those cold and timid souls who know neither victory nor defeat."

- Theodore Roosevelt, from the "Citizenship in a Republic" speech at Sorbonne, France 1910

The way to get started is to quit talking and begin doing.

- Walt Disney

Chapter 19 – Extras

Here are a few extras that we thought you might find useful.

- Contract with the Willing
- Survey Results
- Safety Mantras
- Master Take-Away Points
- An Invitation

Today's Best Practices Contract with the Willing

1. I will embody "Safety First, Safety Always" every day in pursuit of safety perfection.
2. I will identify, report, and resolve safety contradictions.
3. I will advance my learning about safety through training and certifications.
4. I will use my problem-solving skills to improve the safe-vs.-fast contradiction.
5. I will be precisely negative, when appropriate, to make solutions practical.
6. I will understand what I need to do each day to work safe.
7. I will identify and prevent hazards from happening.
8. I will be vigilant to identify changing job conditions affecting everyone's safety.
9. I will support and lead our culture of safety.
10. I will fully accept responsibility for my actions.
11. I will immediately correct safety issues wherever I find them.
12. I will use the correct tool and process for the job.
13. I will care enough to never compromise safety.
14. I will solve problems that advance safety and production.
15. I will never be afraid to ask questions.
16. I will keep an open mind each and every day.
17. I will hold management and myself responsible for a safe workplace.
18. I will contribute to safety in my professional and personal life.
19. I will work safely for my family's and friends' sake.
20. I will do the right thing - even when no one is looking.

By my signature I herewith agree that I am willing and fully committed to take full responsibility for my safety, to embrace prevention, to resolve safety contradictions, and help my co-workers to work safe as well.

Signature of the Willing	Date
Signature of their Supervisor	Date
Signature of their Project Safety Manager	Date

SURVEY RESULTS

Tell me and I forget. Teach me and I remember. Involve me and I learn.

- Benjamin Franklin

After incorporating some of the concepts in this book into trainings, JHAs, etc., we took a survey on a big construction project. We requested worker opinions about safety today and how safety relates to their current work.

The survey contained six questions. About 400 surveys were handed out to craft workers, engineers, and supervisory personnel. Surveys were returned and the results were collated. Sample size was statistically significant and randomization was not considered so the results may not apply to the entire jobsite or to all jobsites. The comments were made free-form. The data was organized into categories to provide basic insight into the greater meaning of the response. No root cause analysis was conducted.

The reason we are incorporating the results here is so you can see what craft workers think about safety while working on a site that incorporates today's best practices. It also shows what workers believe can be done to make improvements after they have received some of the training suggested in this book. We believe this survey demonstrates that the analysis of today's As-Is condition is pretty close to reality from a craft worker perspective. The survey also demonstrates: craft workers want improved safety, they will incorporate important concepts into their mindset, and today's construction mindset is not "set" in concrete. Finally, we are heartened to see that even with minimal training, the craft workers are open to these ideas and are willing to dedicate themselves to improve safety. Below are the results of the survey.

Question #1: Why has the project had so many incidents?

- HUMAN ERROR: 32%
  - Complacency, loss of focus.
  - Supervision not doing a good job.
  - Long commute time to jobsite, lack of sleep.
  - Nature of the business.
  - Low morale.
  - Material storage too close to heavy equipment.
- INEXPERIENCE: 29%
  - Training too many craft people at once.
  - Not paying attention to detail, not thinking ahead.
  - Lack of proper training and safety enforcement.

  - Total focus and total engagement required for this type of work.
  - Lack of professional talent to build the work.
  - Overwhelmed at the different safety procedures involved for each task
  - Not physically fit or competent to perform the work correctly the first time.
  - Reporting all incidents.
  - Subcontractors not held to the same safety standards as the prime contractor.
- SCHEDULING: 27%
  - Working in a rush and not understanding the plan.
  - Constant schedule changes and then rework.
  - Not enough planning ahead by management.
  - Production over safety.
  - Being spread too thin; too easy to cut corners and take short-cuts.
  - We need to slow down, stop and view surroundings and situations.
  - Working long hours to meet deadlines.
- WEATHER RELATED ISSUES: 8%
  - Adverse conditions, equipment and footing slippery when wet.
  - Weather related gear restricts movement, visibility, and slows down reaction time.
- LACK OF COMMUNICATION: 4%
  - Lack of communication between trades.
  - Field crews are left to "figure it out on their own".
  - New-hires should have a project tour so they will feel a sense of pride and involvement.

Question #2: What needs to happen in order to eliminate these types of incidents?

- MORE SUPERVISION: 56%
  - Keep people aware of their surroundings.
  - Prevent time wasted, have tools and materials available for crews before they start the work.
  - Hire competent people; need more intense screening.
  - Need more "Hands-On" training for new-hires.
  - Superintendent needs to spend more time working in the field with foremen.

  - Stricter accountability in regards to termination for violating company policies.
  - Follow JHA procedures, review before each task.
  - New eyes needed on Safety walk-through.
  - Find ways to build morale.
  - Reassure people that "*THEIR*" safety is first and foremost.
  - Simplify work plans, give clearer directions.
  - Use smaller crews that work in smaller areas instead of spreading the workers too thin.
  - Consider revising work plans that put people at risk to accomplish the task.
  - New Engineers need to work more closely with seasoned Engineers.
  - Verify work plans before they are sent out for production.
  - The foreman needs to communicate the level of danger, large or small, before work starts.
  - Need a quicker response to questions.
  - Pay more attention to a foreman's request to lay a person off when they don't have safety buy-in.
  - Doing it right the first time does not mean a waste of time.
  - Stop cutting corners to push production.
  - Provide crews with current drawings; clearly communicate what the daily goal is, and location.
  - Treat people like adults; make sure safety literature is adult-based, not politically correct living advice.
  - If any work situation puts your crew in harm's way, there is probably a better way to do it.
  - Make sure all crews get information when an incident occurs, not only the foremen, so that everyone can learn from what happened.
- MORE ACCOUNTABILITY: 37%
  - Craft workers should be responsible for their own Safety.
  - Clean-up work areas and take more PRIDE in your work.
  - Pay attention to your surroundings and don't take short-cuts.
  - Listen to other people, focus and be alert.
  - Foremen need to take more responsibility.
  - Speak-up when a condition does not seem right.
  - Look-out for one another; respect each other.
  - Don't let the stress of the production schedule affect judgment; Safety First, Production Second.

  - If people keep making the same mistakes, are they trainable?
  - Are you working too many hours?
  - Watch attitude and morale.
- BETTER EQUIPMENT: 6%
  - Fix unsafe conditions; need better lighting.
  - Chokers should have a sliding hook set-up.
  - Focus on major safety hazards such as adequate anchor points for our retractable yo-yos (lifelines).
  - Horns on cranes, some not loud enough.

Question #3: Why do you think people take shortcuts or put themselves at risk?

- SELF INFLICTED: 45%
  - Laziness.
  - Learned bad habits on other jobs.
  - Complacency.
  - Inexperience or lack of training.
  - Worried about underachieving.
  - Not thinking.
  - Don't care about safety.
  - Don't listen.
  - Easier.
  - Underestimating the danger of the task.
- PRODUCTION SCHEDULE: 29%
  - Pushing to meet deadlines and taking a risk to get the job done.
  - No accountability and lack of reinforcement from superintendents.
  - Don't want to disappoint their supervisor.
  - Schedule pressures.
  - Consider shortcuts due to loss of production.
- WORKING TOO FAST: 23%
  - Want to impress management.
  - Competition.
  - Feel that "time is money".
  - Feel rushed.
  - Finish sooner.
  - In a hurry to complete a task, and feel there is not enough time.

  - Pressure from foreman.
  - Worried about job security.
- POOR PLANNING: 3%
  - Proper tools, safety gear, or materials are not easily available.
  - Not enough supervision.

Question #4: Do you believe we can prevent all incidents and if so how?

- NO: 52%
  - Too hard with the amount of man hours.
  - You can never eliminate human error; and no one is perfect.
  - You can only minimize them, but the company is on the right track.
  - We could prevent all major ones but not all incidents.
  - We need better planning and no exceptions on safety violations.
  - Getting 200+ people every day to be free of incidents is not realistic, so we need to select the areas with the greatest risk and take steps to not let anyone get hurt.
  - As long as we require every single bee sting, hurt feelings and headaches to be turned in, we will be subject to reported incidents.
  - New-hires are always at risk until they become aware of site operation and their assigned task.
  - Workers will always try to do jobs faster.
  - Getting past the first big milestone will help slow down the injury factor.
- YES: 46%
  - With stronger accountability given to the individuals that do not comply.
  - By preparing detailed JHAs and revising them when necessary.
  - Management needs to increase presence in the field.
  - Making sure all employees are trained on an individual basis; they should not be pushed hard enough to feel they need to take short-cuts.
  - Safety has to be first in everything we do.
  - Don't take short cuts; do things right the first time.
  - Be dedicated and aware.
  - Slow down and plan your next move.
  - Be involved and watch our work; foremen need to look at each person and evaluate whether they are capable of the task.

- It requires more shut downs, communications, and foremen with smaller crews.
- Complete focus on the task at hand.
- Fully trained people.
- Craft should be responsible for their own safety and everyone needs to be more accountable.
- Develop a new mind set; strengthen belief system that incidents are preventable.
- Get rid of complacent people.
- Starts with planning, understanding the operation, and making expectations clear.
- Use the right tools for the job; take pride in what you do.
- Correct all unsafe acts.
- It takes a total commitment on all levels; set realistic goals.
- We need to think safety first, then quality and lastly production.

Question #5: Do we need to do something different as a solution to stop incidents from happening on the project?

- STRONGER COMMITMENTS: 60%
    - All of us can teach, learn, and listen.
    - More one-on-one management talking to craft people.
    - Request more input from crews.
    - Standing back and watching work; work smarter.
    - All forms of management need to engage with different people daily; get to know people better.
    - Planning your work, working your plan.
    - Less stress on production schedule, and more focus on doing work right the first time.
    - Eliminate rework! It increases the odds of incidents.
    - Making sure all employees are trained.
    - Being more involved in your work.
    - Foremen need to believe in the motto, "Nobody Gets Hurt", and expect it to happen.
    - Getting a 100% commitment from everyone.
    - Believe that you can always do better.
    - Communicate clearly; narrow plans for the day and take small chunks of work and improve them.
    - Hire more qualified people.

- Better planning of supplies/materials and logistics will stop short-cuts.
- Superintendents need to be responsible to see that 100% of the task is completed before moving on.
- Not relying on second-hand information - physically watch the work being done yourself.

- NO or WAS LEFT UNANSWERED: 19%
    - Incidents happen, human nature.
    - No, I think all people are trying to work safe.
- FRESH SAFETY FOCUS: 14%
    - Keep changing things up so that the safety emphasis is fresh and interesting.
    - Good lead (experienced) people should be paired-up with less experienced people.
    - More safety recognition.
    - Keep discussing safety from large to small acts.
    - Designate small crews to check safety procedures.
    - Appoint a crew safety leader with special training to become the voice of the crew for a given period of time.
    - Acknowledge a no-incidents record.
    - More "policing" policies and continue stressing safety.
    - Treat crew like adults; treat incidents with common sense, and the numbers will drop.
- MORE ACCOUNTABILITY: 6%
    - Stronger accountability given to individuals that do not comply with safety rules.
    - Allow more flexibility (have a say) to foreman to let a person go (terminate) when someone seriously violates or is not bought in to best safety practices.
    - Don't leave a mess; cleanliness is a safety issue; keep up on housekeeping.

Question #6: What more can I do to keep my crew safe each and every day?

- USE MY EYES MORE (ANTICIPATE): 29%
    - Keep a key eye out for my crew and surroundings.
    - Be more vigilant.
    - Pay attention to the way a person works, know their attributes and limits.

  - Look after each other (crew members).
  - Know the people you are working with and know their capabilities, watch them work and make suggestions.
  - Pull my crew out of unsafe or overly congested areas even if it slows down production.
  - Never walk by an unsafe act and don't be involved in one.
  - Look further ahead in my operation.
  - Follow safety rules and watch for changing conditions.
  - If any work situation puts crews in harms way, there is probably a better way to do it.
  - Continue to help identify hazards.
- USE MY HEAD MORE (THINK): 29%
  - Pay attention and always keep safety in mind.
  - Keep focused on safety first every day.
  - Be more detailed. Be better prepared.
  - Stay focused and lookout for each other.
  - Try to minimize overtime or at best rotate to eliminate burnout. Get people to appreciate our safety culture and buy into it.
  - Provide the correct tools and information.
  - Make sure everyone has what they need to be safe, and keep safety supplies stored-up.
  - Have better work plans. Be better prepared.
  - Keep JHAs current and make changes to them when conditions change or new hazards are identified.
  - Use proper tools for the job.
  - "Awarding/recognition".
  - Hands-on training. Better housekeeping.
  - Be involved with work and set-up for safety success during planning.
  - When you are on top of your game, there is less finger pointing and more results.
- USE MY VOICE MORE (COMMUNICATE): 29%
  - Remind them daily.
  - Speak-up. Train my people more.
  - Commit to good communication.
  - Continue to talk and teach safety.
  - Communicate what and why for tasks.
  - When revising a JHA, ask for input from all people present.

- Try to give them the information they need and talk about what could happen so risks are in everyone's mind.
- Make sure they know your expectations: That they take the time to do their work safely and with the highest quality.
- Always bring up whatever comes to mind. It does not matter the time of day. Do not let moments pass where safety can be communicated. Even random thoughts can matter – Speak Up.
- Talk with my crew and give them the confidence to do the job correctly the first time, with the time and knowledge they need.
- Continue to emphasize that complacency is not tolerated.

- USE MY HEART MORE (CONNECT): 7%
  - Keep up with housekeeping, water removal (these guys work hard and deserve to not work in ankle deep water all day).
  - Allow them to have fun and not be so up-tight. Teach them and let them have their say.
  - Give them praise when deserved and correct behavior when needed.
  - Be a TEAM (Together Everyone Achieves More).
  - Assist in the operation of a struggling task.
  - Be there with them; be involved in the work.
  - Respect each other and their work schedules. Attempt to work in areas that are not congested.
  - Get to know my crew, and then express commitment to safety, addressing all concerns, even the small ones, builds trust.
- USE MY EARS (LISTEN) MORE: 6%
  - Listen to concerns of the crew.
  - Listen-up. Talk safety everyday; never be afraid to speak up.
  - Each person on the crew has something special to offer, listen, acknowledge, and give credit.
  - Keep crews involved, enlist their input, lead them through clear thought – get them to "out" any situations.
  - Follow through with answers to questions.

Summary

If you read the data carefully you will see that it supports the many issues exposed in the current situation analysis and discussions we had in Part One.

- Craft workers perceived schedule pressure to be against working safely.
- The complexity of the work and inexperience are important safety issues.

- Greater management involvement and supervision is perceived as important to safety success.
- Personal accountability to oneself and peers is perceived as important to safety success.

This is not really new, is it?

But did it surprise you how many people thought that incidents could be completely eliminated (46%)? This was a remarkable finding and not expected. This tells us that craft workers in general are more optimistic about the possibility of excellent safety outcomes despite the barriers to safety success that they face and intense schedule pressure.

60% made a comment related to having a stronger commitment to assure safety success. Based on the many comments about planning, material and work issues, it is clear that processes are not standardized or well structured. There is incredible opportunity to make improvements in safety, in productivity, in process, and in quality for construction.

14% wanted a fresh focus on safety, which is a form of increased commitment. Commitment matters. That should come as no surprise.

When it came to a shared safety commitment craft workers felt they needed to anticipate risks better, think before they took action, and communicate to each other about risks, which shows they realize they are not acting on all they know. This strengthens our resolve to eliminate the hidden drivers that cause workers to act wrongly.

What would your craft people say about the jobsite where you work?

> *"An honest safety dialogue begins with listening in order to understand worker opinions." - Brion K. Hanks*

Safety Mantras

Safety Mantra Introduction: Nobody gets hurt near me.

Safety Mantra #1: Safety is always under construction.

Safety Mantra #2: Risk comes from not knowing what you are doing.

Safety Mantra #3: Safety solutions must be win-win.

Safety Mantra #4: Negative emotion and safety are like oil and water; they don't mix.

Safety Mantra #5: Safety begins with me and my attitude.

Safety Mantra #6: Win-win is created when we solve tough safety problems together.

Safety Mantra #7: Safety contradictions are opportunities.

Safety Mantra #8: Your safety record reflects your integrity and your knowledge.

Safety Mantra #9: A safety culture is made moment-by-moment and choice-by-choice.

Safety Mantra #10: Choose well. Actions have consequences.

Safety Mantra #11: Learn from the masters to create your safety foundation.

Safety Mantra #12: What you don't know hurts and kills people.

Safety Mantra #13: One for all, and all for one in safety.

Safety Mantra #14: Expect the unexpected and prepare for it now.

Safety Mantra #15: Safe work demands sustained attention.

Safety Mantra #16: People do weird things when they are afraid and unprepared.

Safety Mantra #17: Safety perfection is a journey of safety mastery.

THE BOTTOM LINE: SAFETY UNDER CONSTRUCTION

Introduction - Why Write About Safety?
Bottom line Introduction: People continue to get hurt and killed on the job. The new safety goal is 100% zero harm.

Chapter 1 - Safety Defined
Bottom line for Chapter 1: If we are fully responsible and face our fear and pain, we can realize there are no accidents. There is only what we do not know today, the errors we currently make, the consequences of our errors, and what we will know tomorrow.

Chapter 2 - Why You Are Not Safe Yet
Bottom line for Chapter 2: You are not yet safe on a construction project, because your safety knowledge is insufficient to produce that goal reliably.

Chapter 3 - What's Coming and How to Prepare Yourself
Bottom line for Chapter 3: Processes and projects will have safety and quality built-in, with cost and performance as prime considerations and meaningful work for all.

Chapter 4 - Analysis of Safety Flaws and Failures
Bottom line for Chapter 4: "Accidents" occur near hidden mindsets, denial, high emotions, indifference, compromise, or suppression of the truth.

Chapter 5 - Current Best Practices of Attitudes
Bottom line for Chapter 5: Character and leadership are the ultimate expressions of attitudes that make you safe.

Chapter 6 - What We Can Learn From the History of OSHA
Bottom line for Chapter 6: New levels of safety and profit performance will be achieved through mutually engaged stakeholders striving for win-win solutions, resolving contradictions, and using a multi-disciplinary and scientific approach to safety.

Chapter 7 - Safety Contradictions / No Compromise
Bottom line for Chapter 7: Construction production and safety contradictions are opportunities disguised by your fearful mindset - a belief that you cannot solve tough problems.

Chapter 8 - Today's Minimum Safety Practices
Bottom line for Chapter 8: If your best practices are achieving 95% safety on the job then your priorities are 95% correct. Remember that when you stake someone's life on a countermeasure you must have 100% vigilance and multiple backup plans. To get that last 5% your priorities demand, you must move beyond compromise and countermeasures.

Chapter 9 - The Current Culture of Safety
Bottom line for Chapter 9: Build a culture of action based on safety priorities. Use best practices but go further. Continuous safety improvement is mandatory for survival.

Chapter 10 - Responsibility and Consequences
Bottom line for Chapter 10: Accept heartfelt responsibility for safety because positive safety consequences come from positive choices.

Chapter 11 - Best Practices Virtual Walk-Through
Bottom line for Chapter 11: Live safety best practices. Focus on them. Dedicate yourself. Embody excellent safety actions and choices in each moment.

Chapter 12 - Ignorant Things We Say and Believe
Bottom line for Chapter 12: A hidden mindset of risk and recklessness leaves a trail. Look for evidence in the form of ignorance, assumptions, excuses, unclear thinking, glib comments, Russian roulette, and false learning.

Chapter 13 - Examine Your Thinking Patterns
Bottom line for Chapter 13: Examine your thinking patterns. Be responsible for your mindset inertia. Connect to your crew but be your own person standing up for the highest safety ideals.

Chapter 14 - Outcomes of Split-Second Decisions
Bottom line for Chapter 14: Split-second decisions happen in an instant but it takes intensive preparation to be ready for those moments where safety hangs in the balance.

Chapter 15 - The Complacency/High-Skill Contradiction
Bottom line for Chapter 15: Reduce complacency by keeping yourself engaged and attentive to the task at hand.

Chapter 16 - An Inventive View of a Culture of Safety
Bottom line for Chapter 16: Stress, fear, uncertainty, and awkward social situations drive unusual automatic behaviors and create fuzzy thinking in people. If we are to achieve 100% zero harm we must understand and resolve these underlying problems.

Chapter 17 - Safety Leader Behaviors, Mindsets and Character
Bottom line for Chapter 17: Safety perfection begins with a problem-solving leader who accepts responsibility, resolves contradictions, and persists until both a 100% zero harm environment and productive processes are delivered. This leader must have a process focus and a systems viewpoint. Learning is now a part of the new safety and construction mindset because continuous safety improvement is mandatory for survival.

An Invitation

We invite you to contact us if:

- You have an extremely thorny or unique safety challenge.
- You do not know where to start or what to tackle next.
- You have a great story or experience to share.
- You think we missed something important.
- You wish to express some precise negativity. We're listening.
- You are dedicated to promoting a zero harm safety environment.

Whether you are an individual, regulator, union member, company executive or safety leader at any level, we encourage you to send us your real life examples. Send us comments about our book that helped you or angered you or that you flat-out don't believe. Or if there was something we seemed really excited about, but you didn't understand, contact us and ask for an explanation.

We are interested in hearing about your testimonials; especially if you initiated an idea based on reading our book and it turned out to be instrumental in saving a life or was the factor that kept someone from doing something unsafe.

We are also interested to hear what you are doing to improve safety in your organization and on your projects.

Please indicate whether it is okay to post the communication on our website (so we can comment on it) or if it is private. We are strengthening our safety understanding and also building a more relevant safety community. To do this, we'd love to examine real life situations and demonstrated results.

Come visit the book's safety-centric website to access additional resources and contribute to a safety leadership community.

http://SafetyUnderConstruction.com

http://SafetyUnderConstruction.net

Final Author Comments - Brion

Brion K. Hanks, one of the authors of this book, has a passionate dedication to safety. He has over 40 years of experience promoting safety within the construction industry and is very hands-on when it comes to safety every day. His resolute commitment is inspirational to the people that have worked with him. His actions fully embody his teachings. Brion seeks to find new ways to express, teach and improve safety. He is a safety expert and understands the best practices in his field. He has helped to implement many innovations in safety throughout the years. Brion understands that his work enhances the lives of many people. When fathers and mothers safely return home to children and spouses and that seems ordinary, he realizes the profound nature of his work. To say "Brion cares" is a complete underestimate of who he is and what he is about.

Part of Brion's reward for his safety efforts has been the gratitude expressed by hundreds of people who have personally sought Brion out to thank him for helping to keep them safe; for getting to know them on a personal level; for teaching them ways to work safer and for truly caring about their welfare.

> *"Thoughts without content are empty, intuitions without concepts are blind."*
> *- Immanuel Kant*

A Starfish Story—Making a Difference

This is a classic story told to Brion in 1994. I wrote my version adapted from the original work of Loren Corey Eisley, *The Star Thrower* (1978).

Once there was a great storm where the land meets the sea. Amongst other things, washed up on shore were hundreds and hundreds and hundreds of starfish. At dawn a young boy was observed picking up starfish and returning them to the sea. An old man walking the beach had been watching the boy picking up starfish one after another. Finally he approached the boy and said, "Why bother, the sun is coming up and they all are going to die anyway so what difference does it make?"

"What difference does it make, (as the boy held a starfish in his hand), it will make a difference to this one," and he turned to set the star fish free.

My takeaway from this story is that we all can make a difference both in our professional and personal lives. Even with doing little things we can make a difference. We may not realize that a "little thing" we did made a difference to someone or something. It is in the doing, regardless of how big or how much, that a difference gets made.

Having said that, I want to tie the Starfish Story to making a difference regarding safety in the workplace; a place where *you* can make a difference by speaking up, by listening up and then by acting on concerns brought to your attention. You can make a difference by active participation in the development of an operation JHA. You can make a difference by never walking past an unsafe act or condition (This is Kiewit's First Rule of Safety). You can make a difference by always calling a co-worker out if they are doing something unsafe. You can make a difference, when doubtful about something, by stopping to assess or discuss a situation or changed condition. You can make a difference by never taking a short cut to get something done quicker. You can make a difference by getting tool training and you can make a difference by doing the right thing, even when no one is looking (you will know and your personal integrity will be intact). I'm sure there are more things you will think of as to how you can make a difference.

To make a difference in work environments means you work safely and, within your sphere of influence, do everything within your power to help keep co-workers safe.

How about this for at home safety and making a difference? A worker on a recent project (2012) personally conveyed to me that he was so impressed with Kiewit's safety program and the daily 'Play of The Day' that he started a daily program at home with his seven children. He wanted them to think about hazards associated with everything they did and it was working to the point where he became amazed that his very young children were pointing things out and discussing them with their Dad. I told him I was proud of his efforts to apply Kiewit's safety programs at home with his family... that his children will undoubtedly rub off on other children and so they too will be making a difference. This is a great and positive example of how the Kiewit safety ripple-affect grows.

So, a question of great consequence needs to be asked of you, "Are you making a difference?" It's a simple question!

Whose Funeral Is It?

This story was written by Brion K. Hanks in 2013.

Like a lost soul wandering in the wilderness, you've been confused, walking around in circles near a cemetery and wondering why you feel so lost. You observe that a funeral is taking place. Something is drawing you into the cemetery. Now you are approaching the funeral ceremony and you can see and hear people crying and consoling each other. You say, "Whose funeral is this? I recognize people here."

Approaching the open casket, you first see your Mom and Dad, your pregnant wife and other family members who are grief-stricken. The person performing the funeral ceremony is heard to say, "He was so young... this should not have been his time... to die in a construction accident that should never have happened... such a waste..."

Still wondering who died, you approach and your knees buckle upon seeing yourself lying in the casket. You are in shock! How can this be? What happened?

Here's your answer. Three days before you were on a construction site where a large excavator was digging near a wall. You walked behind the excavator while assuming that it was going to turn the opposite direction from what it did. You failed to communicate with the operator and put him on notice of your intent to move behind the equipment. He never saw you and you were crushed between the wall and the counterweight of the excavator and died right there on the scene.

If you remember, you had been warned on two previous occasions to stop moving around equipment without checking in with the operator first. You committed a serious unsafe act by again ignoring the required safety procedure, and have now paid with your life. Your death was senseless and should never have happened.

The tragedy in this story is that something similar is played out in real life far, far too often. Let this be a sobering moment in your life where you, the reader, commit to never do such a foolish thing while at work. Further, that you hold people accountable within your sphere of influence.

Now go and live out your life as it is meant to be, and be all that you can by making continual safety differences.

Final

To be relentless in pursuit of Safety Perfection™ is to be on the right road to keep this noble goal always at the forefront for everyone to see. My hope is that you will passionately engage and fully participate in helping to build this safer future for the sake of the people within your organization. Responsibility coupled with your passionate will and continuous competency can create safety buy-in and can become a powerful dynamic rewarded with safety success. Safety is always under construction on your path toward Safety Perfection. To that end, I wish you enough!

FINAL AUTHOR COMMENTS - INNOVATION SCIENTIFIC

SCOTT BURR AND DAYNA HUBENTHAL

Terminology and Conventions

We use terms throughout the book such as “innovator” or “professional innovator” or “problem-solver” or “inventor”. When we use these or similar terms we mean that this person has applied an essential principle of structured innovation. When we use the terms from an “innovator’s view” or an “inventive viewpoint” we are discussing how changes in perspective or that overcoming psychological inertia has occurred.

Various methods, practices, and processes comprise the structured innovation system. When all of those ‘tools’ are mentioned en masse, the authors refer to “the Structured Innovation toolset” or the “Structured Innovation system”. This Structured Innovation system is based on TRIZ and adapted to accommodate the more impatient Western mindset. It redefines some tools, adds new ones, and eliminates the use of other tools. TRIZ strives for a perfect solution whereas a Western problem-solver strives for effective solutions within a context – as good as needed is good enough, but never defaulting to a compromise!

TRIZ practitioners deliver highly effective results but can sometimes be out of step with U.S. management priorities. The strength and weakness of both TRIZ and Structured Innovation is the psychological inertia of the person applying it. We are grateful to Boris Zlotin, Alla Zsuman and Dana Clarke for their pioneering viewpoints and service to the scientific community.

At Innovation Scientific, it is seldom necessary to make use of the full TRIZ or even Structured Innovation toolset to create a breakthrough. Most innovations can be achieved using a small selection of “tools”; therefore, when the authors reference a part of the system, we will call out that principle, section, tool, or process by name and explain it.

Research Methods

We use qualitative research to gather mainly verbal/observed data and analyze the information in an interpretative manner (subjective, impressionistic and diagnostic). We try to understand the information from a social, communication, cultural and historic (anthropological) context whenever possible. We almost always work with subject matter experts as participants to understand peoples’ behavior, opinions and decision-making processes whenever possible.

We also perform quantitative research, which generates numerical data or information that can be converted into statistics, trends, and numbers.

We make heavy use of the web by using semantic and proprietary search methods. The trained human brain is still far superior for this purpose. It cannot yet be matched by even the most advanced semantic search engines.

The Challenge of Proper Citations for an Inventive Process

Because we use such a wide variety of research techniques and use several unrelated disciplines of science combined with the innovation process, we sometimes come up with our own theories and confirm them without realizing that other experts have worked on the same topic before us but under a different field of study.

In general we find that the whole system of "citation" of previous fundamental works to be very challenging for a multi-disciplinary approach that is inventive and not fully scientific. [69]

Innovators are usually experts in only a few topics and so may not know the pioneers of other fields of study. We often "reinvent the wheel" because it is easy and inexpensive for us to do. However, when we find something that is very fundamental and highly respected, we source these works even if we did not use them. Occasionally, we do not find a source that should be cited, however; so if we have stated something in this book and you believe we should cite a supporting source, please let us know. We love to have multiple sources for ideas; this creates strong foundations.

Innovation and the Science behind This Work

Having a toolset is great but a master craftsperson knows that having a saw does not mean you know how to use a saw. The same is true of TRIZ and Structured Innovation.

Safety Under Construction for Leaders, Engineers and Innovators (the first in this series - soon to be published) documents the outcomes of the process Innovation Scientific used with Brion K. Hanks to create this body of knowledge. This book is more rigorous in general and the science is explained in greater detail. As you read that book you will experience the process of making the inventive connections yourself.

---

[69] In the long-term we plan to work with other leaders in multi-disciplinary inventive work to create a system of citation that supports citing previous thought leaders.

Any person interested in experiencing the emotional arcs of an innovator and seeing the documentation of an innovation project arranged as a sequential action-learning process would benefit greatly. We especially encourage professional innovators to read it. If you are interested in the science, references, and innovation process used to create this body of knowledge please go to and enjoy the website http://InnovationScientific.com and order the book.

Final Thought

If you find yourself highly motivated by this work please do engage and participate in building upon this paradigm shift. Start with yourself and grow your sphere of influence through positive results. Take responsibility and couple it with your passionate desire to go home every day to the people who love you and who count on you being in their lives. Grow your contradiction-solving competency so you will remain relevant as the new safety future unfolds because safety is always under construction.

Build the safe and profitable future you want by behaving as if it is already here. Be the person you need to be, but start it right now. Involve others. Build a strong safety community. Let us know about your progress. We wish you a fulfilling, healthy and safe life.

Thank you.

www.ingramcontent.com/pod-product-compliance
Lightning Source LLC
LaVergne TN
LVHW062120200926
848484LV00007B/1722
9781938282256